BEATING TRAFFIC

Time to Get Unstuck

Michael L. Sena

Bloomington, IN   authorHOUSE®   Milton Keynes, UK

AuthorHouse™
1663 Liberty Drive, Suite 200
Bloomington, IN 47403
www.authorhouse.com
Phone: 1-800-839-8640

AuthorHouse™ UK Ltd.
500 Avebury Boulevard
Central Milton Keynes, MK9 2BE
www.authorhouse.co.uk
Phone: 08001974150

First published by AuthorHouse 2/5/2007

ISBN: 978-1-4259-6243-2 (sc)

Library of Congress Control Number: 2007900144

Printed in the United States of America
Bloomington, Indiana

This book is printed on acid-free paper.

Dedication

Traffic congestion is not new, but it has gone from being an exceptional event for a few to an everyday affair for many. This book is dedicated to all the people in the world who have ever missed an important appointment, meeting or activity because they were stuck in a traffic jam.

Table of Contents

Introduction
Taking Traffic Congestion Personally

You CAN BEAT THE TRAFFIC CONGESTION PROBLEM. You can take back the week or two of time you lose each year being stuck in traffic, and spend that time on something much more useful and productive. This book is intended to help you do just that, to get you off the traffic treadmill by building up your understanding of the dynamics of traffic congestion. We will look at why, when and how congestion occurs, the part that you play in it, and what you and your family can do to reduce the negative effects of traffic congestion on your lives. We will also look at how to best avoid traffic congestion while you are helping to solve the problems that cause it.

First, we need to put traffic congestion into its proper perspective. It has become a deeply polarizing issue, an "us versus them" dilemma. Whether you are an "us" or a "them" depends on whether or not you own or drive a motorized vehicle. If you live in a densely populated urbanized area and, for economic or other reasons, do not own a car or van or pick-up or sport utility vehicle or anything with four wheels and a motor, you may resent all of those cars and trucks clogging up your streets and causing your bus or trolley to be delayed. If you are one of the large majority of people who do own and drive a car, and you are moving everyday at a snail's pace on increasingly choked-up roads to get on with your normal business and your life, you may have a different view of congestion. You might resent the bus lanes and trolley tracks, the pedestrian crossings, the red lights and stop signs.

Car drivers are being scolded for polluting the environment and wasting precious natural resources. Anti-car groups are forming alliances all over the world to make purchasing, owning and driving a car more difficult and more expensive than it has been in the past so that people will abandon them and flock to public transit, or start walking or cycling as we did before cars and trucks proliferated. These groups are promoting the elimination all cars and trucks from urban areas, are recommending heavy usage taxes for all roadways, and want to make truck transportation so difficult and expensive that goods transport will be forced to return to the rails. Here are a few examples of what is happening:

- The Mayor of Paris created car-free zones by having piles of sand dumped at the entrances to major thoroughfares.

- Every Saturday and Sunday, weather permitting, the main street of the Ginza shopping district in Tokyo along with all connecting streets for one block are closed to vehicular traffic.

- The Mayor of London succeeded in instituting a so-called *congestion charging scheme* in Central London. The charges were increased from the original £5 to £8, a 60% increase, just a year after the scheme was introduced in 2003, and plans were on the board in 2006 to extend its coverage beyond central London to the surrounding boroughs.

- Raising the price of operating a vehicle is another method being used to force people out of their cars and off the roads. This is what is done in Europe. The United Kingdom and Norway, both oil producers, have the highest prices for fuel in Europe, triple the price of a gallon of gas in the U.S., and most of it is tax.

These are not isolated occurrences. Car-free cities, inner city car charging schemes and road tolling are being discussed and implemented everywhere in Europe, in the U.S., in Asia and in the Middle East, all in the name of reducing the negative effects of traffic congestion. The list of traffic congestion offenses includes noise, pollution, crowding out of on street public transportation, impeding emergency vehicles, adding danger to pedestrians and the general inconvenience caused by delays in making daily journeys to work, school, recreation and shopping.

A normal Saturday on the main street in Tokyo's Ginza shopping district.
Photo by Britt Marie Sena

Getting people of all ages out of their cars and walking and exercising more is a very good idea. The world's population has gotten obese while its environment has grown more polluted from car, truck, bus, airplane and all other vehicle fumes. Too much riding in cars, too much TV and PC gazing, and too much fast food are the main reasons for the sorry state of our collective health.[1] We need to reduce harmful emissions before global warming is so far advanced that we cannot stop it. We need to stop killing more than a million people a year, and injuring almost fifty times that number, in vehicle accidents all over the world.

Most people agree that we do need to reduce traffic congestion and promote more healthy and environmentally sustainable transportation. Where the disagreement occurs is with the methods that are being promoted by the anti-car lobby to achieve the desired results. Greenhouse gas emissions from cars and trucks are not the principal culprit behind global warming. Transportation, including cars and trucks, but also planes and trains, boats and buses, are tied in fourth place with *agriculture*, at 13.5%, and just slightly ahead of *other*, with 12.9%. The main contributor to greenhouse gases is *electricity generation and heating*, at 24.5%, which is

3

still concentrated on the use of coal as a fuel. In second place is *deforestation*, at 18.2%. In last place, at 3.6%, is the disposal of waste products.[2]

Indiscriminate road closings, road user tolls, high fuel taxes and similar measures punish the victims of traffic congestion, not those who created the original conditions for it and who continue to foster these conditions. Anti-car solutions attack the symptoms, but totally ignore the real causes of traffic congestion, which are, on the one hand, a lack of forethought by governments and planners to build city regions that do not promote congestion-causing movement, and on the other hand, well-contrived business decisions—backed by political policies, legislation and financial incentives—that have allowed urban regions all over the world to develop in ways that make non-car solutions to transportation ineffective and obsolete.

Figure 1: Means of Transportation to Work: 1990 and 2000

Means of Transportation	1990 in%	2000 in%	Change%
Car, truck or van	86.5	87.9	1.3
Drove alone	73.2	75.7	2.5
Carpooled	13.4	12.2	-1.2
Bus	3.0	2.5	-0.5
Streetcar or trolley	0.1	0.1	-
Subway or elevated	1.5	1.5	-
Railroad	0.5	0.5	-
Ferryboat	-	-	-
Taxicab	0.2	0.2	
Motorcycle	0.2	0.1	-0.1
Bicycle	0.4	0.4	-
Walked	3.9	2.9	-1.0
Other means	0.7	0.7	-
Worked at home	3.0	3.3	0.3

Source: U.S. Census Bureau, 1990 Census Summary Tape File 3 and Census 2000 Summary File 3

Most people who are not fortunate enough to live in a city where they can also work, shop, recreate and educate themselves and their children— and where there is actually an alternative to driving, like an operating public transit system—need their cars today to get themselves to wherever they have to go. In most cases, there are no alternatives, no bus or tram or trolley or underground services. In those places where alternatives do exist, they are often significantly inferior to the automobile transportation option in one or more ways. They take much more time, are less convenient from a schedule point of view, are much less comfortable, and often are, or appear to be, less safe than the car option. What else would explain the situation shown in the tables above, *Means of Transportation to Work: 1990 and 2000*?

I do not believe that 87.9% of the American people in 2000 who drove to work rather than taking other means, as shown in the table above, were insensitive about the problems related to car use, or that they were incapable of making choices in their use of transportation that were both rational and in the public interest, rather than in just their own selfish interests. The data in the table shows that fewer persons took the bus than who walked to work! In early 2006, as the effects of multiple shocks took hold from uncertain oil supplies, increased demand for oil products from China to feed their fast-paced growth, and massive damage to domestic oil production caused by hurricane Katrina, pump prices for fuel reached levels previously unheard of in the U.S., $3.00 or more per gallon. Many cities reported increased transit ridership. The American Public Transport Association (APTA) reported that users of public transport rose slightly more than 4% in the first quarter of 2006 versus a year earlier. But 4% of a total of only 5% is still an insignificant number using public transport compared to those using cars, and it is questionable whether these increases are sustainable if the price of fuel falls to previous levels.

There must be a good reason to explain why so many people drive rather than take public transport, walk or cycle to wherever they have to go. There is. People are just trying to do the best they can to use the hours that a day offers in the most beneficial way for themselves and their families. Moreover, stop-gap measures, such as charging vehicles for entering certain districts, or road usage fees, will generate their own secondary effects that could well be more damaging for our city regions and their inhabitants than their currently clogged transportation arteries and capillaries. With entrance fees, it is more than likely that the problem, along with businesses and jobs, will simply be moved to another location where the tolls or restrictions are lower. There is ample evidence that I will present in this book that when businesses feel conditions are unfavorable in one place, they move. When labor rates become too high; when space becomes too scarce or expensive; when crime rates increase; when taxes reach a painful point: companies will move. If businesses feel that their customers and employees would feel better about visiting them and working at their premises if those premises were somewhere else, they will move.

Initial results during the first few months of the congestion charging scheme in London were all positive. Traffic was supposedly down by 30%, with 65,000 fewer vehicles entering the charging zone. *Transport for London* was claiming that the large majority of these people had simply switched over to public transport. But eighteen months later, The *London Chamber of Commerce and Industry* published their Third Survey of the effect on the retail sector painting a very different picture.[3]

- 63% of respondents reported a fall in customer numbers since the charge was introduced;

- 37% of retailers have reduced their staffing;

- 33% are considering relocating to a site outside of the charging zone; and,

- 28% are considering closing their business.

It seems that instead of switching to public transport, many people just did not travel into London. The "non-essential trips" made by shoppers, tourists and some business people, were the reason the numbers had fallen, and it appears that these had the greatest effects on the businesses in the charging zone. The fee was increased in mid-2005 from £5 per day to £8 per day, a 60% increase. One reason the increase was planned well before it was instituted was that because more people than projected were staying away, the scheme was not generating enough money to pay for itself. Then the London bombings happened just as the new fee was going into effect, in July 2005, and many who had previously taken public transport went back to their cars.

So people who were now truly anxious about riding the London Underground had to decide whether to pay the 60% increase—if they could afford it—conquer their fears on a daily basis, or look for a job outside of Central London.

With road usage fees, those who are already living on the fringes, both socially and financially, will have a more difficult time getting themselves to work. This issue is trivialized by economists and environmental activists alike. For example, an article in the New Statesman made the following claim:

"Road pricing will penalise the poor, it will be argued (largely by politicians and journalists who don't usually give a fig for the poor). But most of the very poor do not own cars and those that (sic) *do use them far less than average."*[4]

Part of the reason that the poor <u>are</u> poor is that they cannot get to where the jobs are located, precisely <u>because</u> they cannot afford to own and operate a car. Raising the price of operation makes an already difficult situation for the poor even worse. The author of this article should ask the poor people in New Orleans who were unable to escape the ravages of hurricane Katrina in September, 2005 if they would have preferred to own a car when the warning came to evacuate, rather than depending on taking the bus that never came.

Traffic congestion has no positive effects. It is bad for everyone and everything on the planet. But those who <u>are not fortunate and wealthy enough</u> to afford to live in penthouses during the week where they are close to their jobs, and in villas at the weekend where they are close to their play, <u>should not</u> have to bear the burden of redressing the problem as if they have caused it, and they <u>should definitely not</u> have to feel guilty about being forced to get into their cars, if they are fortunate enough to own one, and struggle with traffic every day of the week.

Congestion on our roads, whether in London or Los Angeles, in Bangkok or Atlanta, is neither a natural nor a predestined state of affairs. It has existed in western, urban regions and Japan for less than half a century, and in other Asian capitals for not more than a few dozen years. Humans caused it, and humans can make it disappear. I truly believe this, and I hope to convince you of this as well. In order to make this happen, the "stuckees" (i.e., you and I) need to become engaged in the traffic congestion debate, and to become part of the solution rather than helpless victims of the problem. We need to take traffic congestion personally.

Even in one of the most heavily congested cities in the world, Tokyo, there are times when the roads are relatively traffic free, like this Sunday morning, when people can get to where they have to go without the burden of traffic. Photo by the author.

What do I mean by taking traffic congestion personally? I do not mean becoming angry or engaging in road rage or other anti-social behavior. People have proven that they can become very irritable and do irrational things when they are stuck in a severe traffic jam. By taking traffic congestion personally I mean that each of us will have to solve the problem for ourselves because there is no one, no government, or business or higher power, who is going to solve it for us. We will need to change our own transportation habits in order to get traffic congestion, in whole or in part, out of our daily lives.

To paraphrase Mark Antony in Shakespeare's *Julius Caesar,* my intentions are neither to praise nor to bury cars and trucks. Long after they are gone, we are likely to remember all of the bad things that cars and trucks brought with them during the time they were the principal means of transportation, while we may well forget most of the good.[5] Cars and trucks are not responsible for traffic congestion, and the people who drive them are not culprits. Motorized vehicles are inventions that can be used for good or evil. Cars and trucks have done a great deal of good in helping us to move ourselves and our goods easier, faster and farther; they have also done a great deal of evil in adding significantly to the pollution of our air and our earth, and to contributing to the death and injury of millions of people since they were introduced one hundred years ago. But they are here, and our lives depend on them. We have to live with them, at least until we make the changes necessary in the way we develop our environment—to right the wrongs of our ancestors—and calling for their elimination is premature.

This book is going to challenge you to think about how you use these inventions, what you and your family do on a daily basis to get from one place to another. First we are going to look at how traffic congestion occurs, its dynamics, and what forces set it in motion. Then we will explore the root causes of traffic congestion, which lie in the way we have built our city regions, planned our transportation systems, and, to a certain extent, the way we have designed our cars and trucks. Next, we will review the types of events that trigger traffic jams and tie-ups in order to begin to discover what we can do to minimize the risks of starting a traffic problem for ourselves and other drivers.

One chapter of the book is devoted solely to a worldwide phenomenon that has had a major impact on reducing available space on already overcrowded roads, namely the increase in truck transportation. Super-sized shopping centers, and their concomitant traffic congestion problems, are the result of this phenomenon, not the cause, and understanding this can play a large part in molding our shopping habits to reduce the impact of traffic congestion. And one chapter addresses the issue of road user charging and alternatives to tolling that may be more effective in reducing congestion and better at achieving a larger number of societal benefits more equitably.

Four specific recommendations are made to help you and your family get unstuck from traffic. They relate to the daily school run, using the car for recreational trips, shopping, and trips to work. Practical steps are presented on how you can change your behavioral patterns to avoid getting into traffic jams in the first place. The final chapter provides a description of tools you can use when you are on the road to see potential congestion dangers well before you are in the middle of them, and to assist you in planning your journeys so that you reach your destination when you had hoped to get there.

My goal with this book is to help you develop a plan that will accentuate the positive experiences of daily travel for you in the future, and, if not eliminate, at least minimize the negative effects of traffic congestion. Knowing the enemy, especially the one within—the one who gets in the car when it would be just as easy to walk—and turning the enemy's weaknesses into your own strengths is the key to overcoming our over dependence on our vehicles and minimizing our risks of landing in a traffic jam. You can change how you get to all the A and B points during a normal day, and you can even change where those points are on your travel map.

You alone cannot eliminate traffic congestion and all the other harmful and damaging side-effects of cars and trucks, but this is one case where the platitude truly fits: By <u>not</u> being part of the problem, you will <u>become</u> part of the solution.

The Dynamics of Traffic Congestion

COLLECTIVE SOLUTIONS TO TRAFFIC CONGESTION, like collective transportation, no longer work in North America and many western European countries. By collective transportation I mean bus, commuter rail, surface and underground fixed rail transit.[6] It is my contention that both collective transportation and organized congestion mitigation ceased working approximately thirty years ago when it appears that transportation academics and practitioners reached the conclusion that traffic congestion was destined to become a way of life. They gave up trying to fix the problem, and began to focus on traffic management. Let us look at the evidence used by these professionals to form their judgments, the fundamental principles of why roads become clogged.

Transportation texts written in the 1960s and early 1970s stated that any road built at that time would be congested at rush hour soon after it was opened. Before this, it was believed that more roads with more lanes and higher speed standards would solve the growing traffic congestion problem. It was reasoned that congestion could not be eliminated, but its effective time could be reduced by building more roads and widening existing ones with more lanes so that the length of the traffic congestion periods would remain relatively constant as the number of vehicles entering the road system increased. Demand for space on the road would continue to grow because of increasing population and transfers from other transportation modes. One author, Anthony Downs, gave a particularly good explanation of the transfer phenomenon in his 1962 paper, *The Law of Peak-Hour Expressway Congestion*.[7] He wrote:

"Recent experience on expressways in large U.S. cities suggests that traffic congestion is here forever. Apparently, no matter how many new super roads are built connecting outlying areas with the downtown business district, auto-driving commuters still move at a crawl during the morning and evening rush hours.

"To many a frustrated commuter, this result indicates abysmally bad foresight by highway planners. However, the real cause of peak-hour congestion is not poor planning, but the operation of traffic equilibrium. In fact, its results are so automatic we can even put them in the form of Downs's Law of Peak-Hour Traffic Congestion."

Commuters, claimed Downs, would attempt to minimize the total amount of time they spent traveling to and from work within the four constraints defined by his Downs's Law. These were:

1. Income – This determines what is economically feasible for an individual. Everyone might like to be taken to work each day in a chauffer-driven limousine, but not everyone can afford this luxury.

2. Money cost of transportation – This includes fares for trains and buses, tolls on highways and bridges, fees for parking and operating costs for a car. The more one is forced to spend on transportation, the less one has available to spend on other necessities and luxuries.

3. Location of residence – Remote, rural areas are unlikely to have public transportation, while central cities may have limited or very expensive parking alternatives. The decision on where one lives has a major bearing on the eventual choice of transportation.

4. Personal comfort – Each person has an individual level of tolerance for different transportation situations, and this can affect the choice of mode to a greater or lesser extent.

Downs went on to define the rules of his Law by classifying commuters into two groups: *Explorers* and *Sheep*. He claimed that most commuters "follow the law of inertia," which means that once they have selected a mode of transportation (i.e., bus, train, foot, car, etc.), they continue to use it until some event encourages or forces them to shift to another mode. Such an event might be a sharp increase in train fares, or the closing of a commuter train line, or the opening of a new expressway, or a park-and-ride facility that offers inexpensive parking. *Explorers* are individuals who are constantly looking for ways to minimize their travel time, and who easily shift modes or routes in order to do so. They are the "rat runners" as they are called in Britain. They can be compared to the shopper who is constantly bargain hunting, or the person who is forever changing telephone subscriptions to get ever-lower rates and bonus payments. *Sheep* are those who rarely shift transportation modes unless a major event occurs.

Models were developed against which the theory of traffic equilibrium was tested. In the early 1960s, metropolitan regions in the U.S. and Europe were still dominated by a central business district (CBD) that was fed by a combination of radial rail lines and roads. Growth spread from the center like the arms on a starfish. Traffic moved like the tide: into the CBD during the morning's rush hour, and out during the evening's exodus. This was before roads like Route 128 and Interstate 495 in the

Boston area, the M25 in Greater London, the *Peripherique* in Paris and many other ring roads around major cities created conditions for multiple centers of business and commerce.

Today, in the 21st century, and during the last two decades of the 20th, traffic flows do not move as they did in the starfish days. Roads that traverse the region connecting multiple nodes are heavily congested in both directions during the peak times, and mildly congested the rest of the time.

However, back then, Downs and others observed that in an environment in which there are only car-driving commuters (i.e., no bus, tram or rail), the building of a new highway connecting a suburban residential area to an urban business district would set in motion route-shifting forces, initially by *Explorers*, and eventually by *Sheep*. Drivers who used the local roads that passed through village centers moved over to the new highway. As the new highway absorbed more cars, the average speed on the highway decreases. Speeds on the local roads increased because there were fewer cars using them because they have moved over to the new highway.

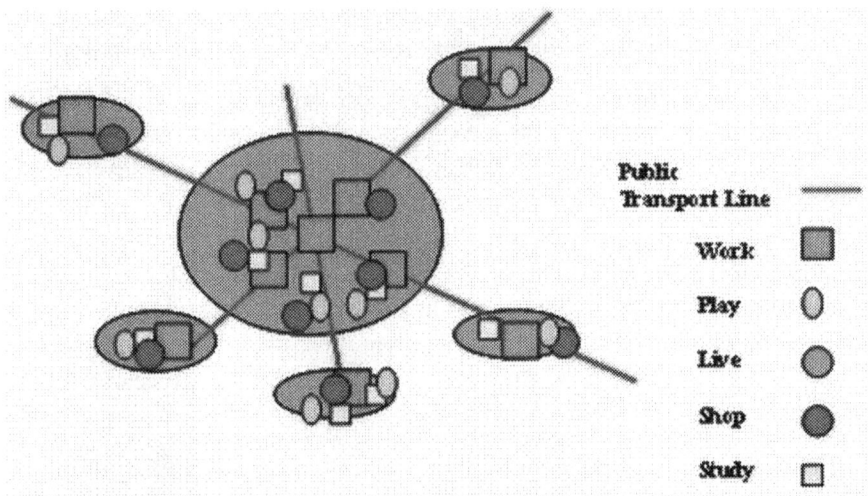

Diagram A: Pre-sprawl
In the pre-suburban sprawl days, transportation connected communities along radial lines to central business districts, where most of the jobs in the region were located. Prior to the rail connections built in the late 1800s, the small communities surrounding the large city were independent, self-sufficient centers of commerce for their surrounding regions. The rail lines converted them into bedroom communities, but most of them retained their commercial, cultural and civic identities.

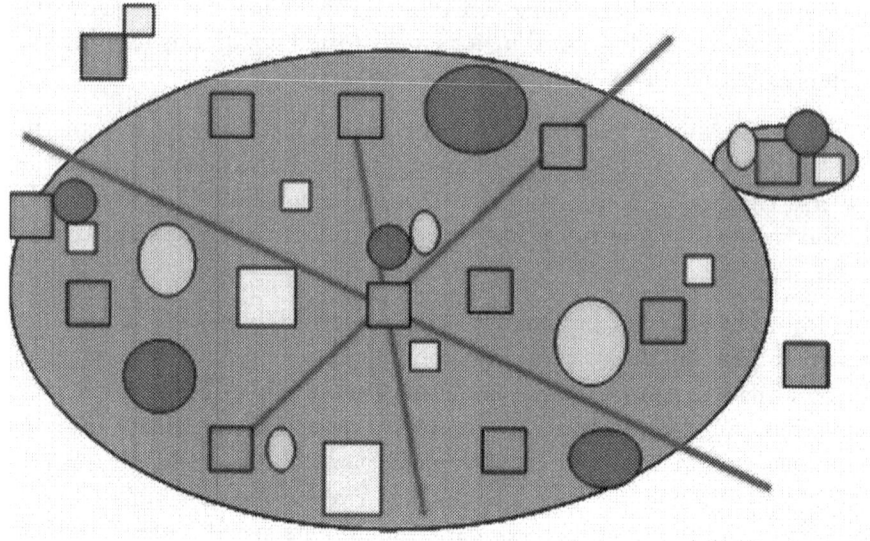

Diagram B: After sprawl

Today, city regions have multiple centers with more jobs located in the suburbs than in what was the central city. Radial transportation networks no longer function, and the car is the only effective means of transport. The former villages have been incorporated into conurbations, and while they may maintain their political independence, they are part of an amorphous sprawl.

In a second scenario, one with a mixed bus and car environment in which new expressways are built to serve the entire region served by bus lines, it was shown that a combination of people leaving the buses to drive their own cars, and car drivers switching from local roads to the new highway, would cause the highway to eventually become congested. As ridership on the buses decreased, the bus lines raised fares and cut service. As a result, more transit riders would become car drivers, and traffic would increase further. Reduced ridership on the buses would set in motion a downward spiral of the bus companies that would eventually lead to their demise.

In a third scenario, when a highway is built alongside a fixed rail line, the transit alternative again loses out to the car. Once the highway is opened, car drivers on local roads switch to the new, faster alternative. Rail riders, initially the *Explorers*, try the highway alternative. The loss of rail ridership has the same negative financial effects on the train lines as on the bus lines, with the added difficulty that off-peak train service cannot compete with the faster, non-congested expressways. Downs predicted (correctly) that the only way rail lines could continue in business would be through heavy public subsidies. A false equilibrium would therefore be established with train lines continuing to operate, but at significantly lower capacity and with heavy government subsidies.

In all these examples, transportation professionals painted a picture that showed how building new, high-speed roads leads to higher congestion than existed before the road openings, and often both higher costs and longer travel times for all travelers. As a result of severely reduced and higher cost services, indomitable transit riders, the last *Sheep*, especially train commuters, would eventually be forced to leave their comfortable seats, where they leisurely read the morning or evening newspaper or enjoyed a chat with their fellow commuters, and join the gathering masses of car drivers. By the early 1970s, many of America's passenger railways had already closed or merged, and their commuter lines had stopped operation.[8] Private bus lines were being taking over by public authorities and operated with grants and tax monies.

What this shows is that millions of individuals making hundreds of millions of totally independent and personally-motivated decisions every minute of every day in every part of the world determine the conditions for every other individual's journey to work, to school, to recreation, to religious services, to shopping facilities, to wherever.

Downs ended his 1962 essay with a challenge and a firm recommendation:

"Any program of expressway planning and construction must be integrated with similar programs concerning other forms of transit in the area if it is not to cause unforeseen and possible deleterious effects upon the level of automobile traffic congestion therein. In particular, marked improvement of roads without any improvement in segregated track transit may cause automobile traffic congestion to get worse instead of better. Since the U.S. has already launched a massive road improvement program, which includes construction of many urban expressways[9], continued failure to undertake an analogous program for other forms of urban commuter transit may result in a generally higher level of rush-hour automobile traffic congestion in those cities which now have extensive segregated track transit facilities serving commuters."[10]

Very little has changed since the behavior of individuals within the transportation infrastructure, and the resulting traffic congestion they cause, were first codified. Anthony Downs refined his *Downs's Law* into his *Principle of Triple Convergence* in 1992 with the publishing of his book, "Stuck in Traffic".[11] This posited that *"traffic congestion occurs because traffic flows in a region's overall transportation networks form almost automatically-self-adjusting relationships among different routes, times and modes."* In other words, build a road, and they will come. People will change the routes they take, switch the times of day during which they travel, and leave the bus, train or car in order to take advantage of a newly opened, faster, cheaper or more comfortable form of travel. He had thirty years of traffic congestion developments from which to draw his new work, and it seemed that his predictions from the early 1960s had become reality.

Ten years after publishing "Stuck in Traffic", as he prepared a second edition of his book, called "Still Stuck in Traffic", Anthony Downs testified before the U.S. Senate's Committee on the Environment and Public Works. His testimony concerned peak-hour traffic congestion. He said he had come to the conclusion that traffic congestion was an unbeatable force, and that there was no better alternative than to accept it. No, not just accept it; embrace it! As they did over forty years ago, his words reflect the most current thinking among transportation professionals, and I have included them verbatim because I feel that it is important to understand this view.

"Most people regard peak-hour traffic congestion as an unmitigated evil," he said to the Committee, *"but that viewpoint is incorrect.* (My underline for emphasis) *Congestion is a vital de facto device we use to ration the scarce space on our roads during periods when too many people want to use that space at once. In effect, congestion is a balancing mechanism that enables us to pursue many other goals besides rapid movement—goals American society values highly. Those goals include having a wide variety of choices about where to live and where to work, working during similar hours so we can interact with each other efficiently, living in low-density settlement patterns, and enjoying highly flexible means of movement—that is, private vehicles. We must use delays from overcrowding in order to pursue the other goals we want to achieve. So congestion makes possible large-scale social benefits as well as the costs of delay on which most people focus when they think about it."[12]*

Similar views have been expressed by other experts in the transportation field. Brian Taylor, director of the Institute of Transportation Studies at UCLA, is quoted in The Economist magazine: "Long queues at restaurants or theatre box-offices are seen as signs of success. Congestion is an inevitable by-product of vibrant, successful cities".[13]

I do not agree with these views, and neither do most people who are subjected to the worst traffic congestion on a daily basis.[14] It is one thing to say that a problem is intractable; it is quite another to say that a problem is in reality a benefit. The "rationing a scarce resource" theme seems like saying that poverty is a de facto device we use to ration the scarce resource of money, or that the plague in the middle ages was a way of rationing the scarce resource of healthy air to breathe in cities. How can the 93 lost, wasted, unpaid hours that a commuter loses <u>per year</u> in Los Angeles, or the 72 hours for a San Francisco commuter, or the 69 hours for a Washington, DC commuter, be a benefit to the individual or to society?[15] Between 1982 and 2003, the average annual hours of delay per traveler for residents of city regions in the U.S. of 3 million or more inhabitants increased by 38 hours.[16] These are hours parents could be spending with their children, doctors spending with their patients, teachers spending with their students. Many workers in the U.S. receive only two weeks of vacation per year. If they could trade in these wasted hours, they could have up to four weeks.

That we all work during similar hours is a gross overstatement. Twenty percent of the U.S. workforce are "shift workers" who work outside the "normal" 9-to-5 period.[17] In the U.K., the figure is an even higher 25%. Non-daytime shift work among health care professions averages 30%; cleaning and building services, machine operators, fabricators and laborers average 27%. The majority of shift workers may not like the hours, but they accept it as a condition of employment. Only 8% of management jobs are outside the daytime hours, and this is presumably where Mr. Downs sees the strongest urge for interaction.

The "congestion is a balancing mechanism" argument is paper thin. Most people today do not have any level of freedom in deciding where they live. They live where they can afford to live. Some families may make certain sacrifices, like not having a second home, or not taking the family for a week every year to Disneyworld, Aspen or Tuscany, in order to live in a community with recognizably better schools. Some families may decide to live in cities amidst the noise, litter and graffiti, and pay higher costs for less space, forego the garden and swimming pool, so that both they and their children can walk to work, school, shops and recreation and spend more time together. But for the majority of people, they put up with living far away from everything, and accept their complete dependence on their cars, because the alternatives are simply not acceptable. I would hardly call that having multiple choices, would you?

As for choosing where and when to work, there was a brief period at the end of the 1990s, at the height of the *dot.com* boom, when it seemed that employees had the upper hand, when employers provided incentives to attract top-flight programmers and program managers: midnight pizza deliveries; free breakfasts and lunches; "play" rooms; free memberships in health clubs and more. That situation didn't last very long. During the last of the eight years that the U.S. had a Democratic President in the

White House, at the height of the boom, Mr. Greenspan cooled down the economy with a series of interest rate hikes. The dot.com boom ended, and so did employee power. Since then, and for most of history, the large majority of people are happy if they have a job that pays a living wage.

Raising the price of owning and operating a car is one of the worst forms of discrimination against the poor. Not having a car or a driver's license is a severe handicap in obtaining and holding a job.[18] Research by the U.S. Federal Reserve Bank shows that people who own cars are more likely to be employed and to work more hours than those who do not own cars; that access to a car shortens periods of unemployment; that car ownership equals increased earnings, especially among racial minorities and low-skilled workers; and, that welfare recipients who received cars through a car ownership program increased their earnings and reduced their dependence on public support payments.[19]

And as for enjoying the "highly flexible means of movement", none of those adjectives fits the private car any longer in most urbanized parts of the world. Driving a car has become a struggle during more hours of the day. In Great Britain, close to 25% of the major arteries are congested for more than one hour per day.[20] It is true that there are no restrictions on anyone getting into their car at any time of the day to travel anywhere they like, but there are no assurances that they will arrive within an acceptable or agreed time, or whether they will arrive at all.

On one point, the proponents of the "traffic is inevitable" message and I do agree: traffic congestion will definitely get much worse before it gets any better. This is true, in part, because of the "feel good, be happy, don't worry" messages that influential professionals are giving to their governments. It's also true because we now have more people living on our planet who have experienced nothing but congestion, and they may have difficulty imagining life without traffic. The message that traffic congestion is with us forever needs to be countered by many other voices spreading the message that traffic congestion can be beaten.

As I will show in the next chapter, to effectively eliminate traffic congestion, we need to get to its roots. If we allow our governments to pursue policies that merely fool us into thinking that congestion is licked because it moves the problem into someone else's community (as with congestion zone charging), or onto someone else's commuting path (as with road tolling schemes), we are only making matters worse, eventually for ourselves—our neighbors can play the same game—but definitely for the following generations.

The Roots of Congestion

The numbers of cars and light trucks[21] in the world have grown from a few thousand at the beginning of the 20th century, to approximately 600,000,000 in 2005. As shown in *Figure 2*, it took fifty years, from 1900 to 1950, to reach 100,000,000. The First and Second World Wars had something to do with the slower growth during the first half century. Production of private vehicles was halted almost completely from 1941 until 1945 in order for car producing countries' factories to support the war efforts in European and the Pacific Theaters. Following WWII, production of cars and trucks expanded rapidly in the U.S., U.K., Germany, Japan, Italy and France.

The number of cars in the world doubled to 200,000,000 from 1950 to 1970, and doubled again to 400,000,000 by 1990. At current rates of sales, it will double again in the twenty years between 1990 and 2010, reaching more than 800,000,000 by 2010. While sales are expected to level off in Europe, where populations are projected to decline during the next three decades, sales in developing countries, such as China, India and Southeast Asia, are expected to more than compensate for the reductions in other parts of the world. It appears that Moore's Law[22] has company in the *Automobile World*.

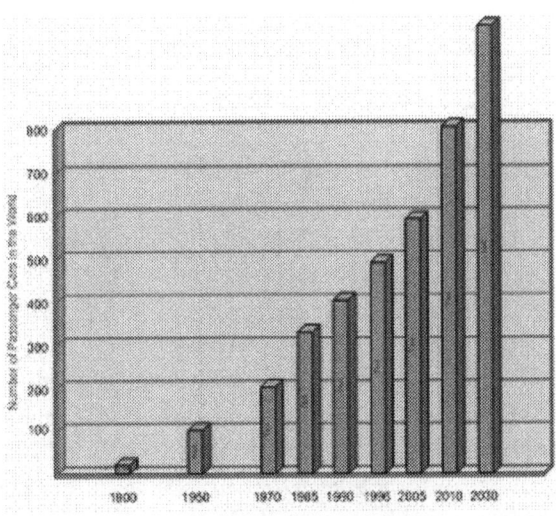

Figure 2: The Growth of Worldwide Passenger Cars in Operation
(Units along y axis are in millions)

Most of North America, much of Europe, Australia, Japan and many parts of Asia are already in the *Automobile World*. Eastern Europe, Russia, China and the rest of Asia and parts of South America and Africa are entering it.

Auto·mobile. It is a word loaded with meaning, both positive and negative. It is what traffic congestion is all about, so if we are going to look at the roots of congestion we should start by tracing the roots of the automobile. An *automobile* is a vehicle, especially one for passengers, carrying its own power-generating and propelling mechanism, for travel on ordinary roads.[23] The word is formed from two individual words, *autos* and *móbilis*. *Autos* is Greek in origin and means "self". *Móbilis* is Latin in origin and means "movable". *Automobile* is "self movable". No horses or mules or oxen or humans required to pull or push the wagon or cart or coach. It moves under its own power. All it requires is the control of a driver, which is a word derived from the Middle English word *driven*, meaning "to cause and guide the movements of".

The word automobile is often shortened to *auto* when used as an adjective in connection with a noun, such as auto sales, or auto repairs. The word *auto* is not usually used on its own. For instance, people don't say, "Let's take the auto out for a spin." In Great Britain, the term *motor car*[24] is more common than the word *automobile*, and in both North America and Great Britain, the word *car* is the everyday, generic term used to refer to automobiles. There seems to have been some borrowing by the Romans from their Celtic neighbors, where *carr* meant "wagon" or "chariot", from which carriage is derived. *Carriage* is a "wheeled vehicle that carries". So *car* is a shortening of *carriage*. In the early days of the automobile, cars were referred to as "horseless carriages".

In Swedish, the language of a country known for automobile manufacturing,[25] the word for *automobile* is quite similar to the English: *automobil*. In Swedish, like its English counterpart, the word is rarely used in everyday speech. The equivalent for *car* in Swedish is *bil*, (pronounced **beel**), which is simply the last part of the word **mo·bil**. Another common Swedish term for a vehicle driven for personal use is *personbil*, which translates roughly into "private car". This is distinguished from *tjänstebil*, which means "service car", or a car that is paid for by one's employer or through one's business. The Germans—never ones to spare letters on words—call a private automobile *personenkraftwagen*, or "personal power vehicle", which is always abbreviated as PKW.[26]

Although the word *automobile* was intended to describe the type of vehicle, it has also come to mean the type of transportation. Automobiles provide personal, as opposed to collective transportation.[27] So it turns out that a*uto* ("self" or "alone") is also appropriate in this context. A large majority of trips in automobiles are made by a single person, the lone driver. In 1999, the U.S. Bureau of the Census reported that 73% of all workers in the United States drove alone to work, while 13% drove with one or more other commuters. A miniscule number, 11%, walked or bicycled or used public transportation, and most of these non-car users lived in or very near large cities like New York City.

The history of the automobile began in the late 18th century. The French engineer Nicolas Joseph Cugnot is credited with building the first self-propelled vehicle in 1789. It was a heavy, three-wheeled, steam-driven carriage with a boiler that projected in front. It lumbered along the streets of Paris at 3 mph (5 kph), slower than a fast walk. In 1801, Richard Trevithick, an English engineer, built a three-wheeled, steam-driven car. Apparently, development of the automobile was held back for decades by over-regulation. Speeds were limited to 4 mph (6.4 kph), and up until 1896 a person was required to walk in front of an automobile carrying a red flag by day and a red lantern by night. Steam-driven vehicles became synonymous with the Stanley Steamer, built by the Stanley brothers of Massachusetts from 1897 until after World War I.

What gave the development of the automobile real acceleration was the introduction of the internal-combustion engine. Karl Benz, a German engineer, is credited with building the first vehicle of this type in 1885. It was a three-wheel contraption. Gottlieb Daimler, another German engineer, built an improved version of the internal combustion engine, also in 1885. In the United States, internal combustion cars were manufactured in the late 1890s by Charles and J. Frank Duryea, Elwood Haynes, Henry Ford, Ransom E. Olds and Alexander Winton. The cars looked like horse-driven carriages without the horses. The early engines had only one cylinder with a bicycle-type drive train. The wheels were wooden carriage wheels. The steering mechanism was a lever.

The automobile industry was almost choked off during its early days because of a patent dispute. The American inventor, George Selden, held a patent that seemed to cover all automobiles. Several companies licensed his patent and formed an association that took over the full rights to the patent in 1907. The association was determined to charge royalties on every car built in America. A group of independent manufacturers, led by Henry

Ford, refused to acknowledge the patent, and took the issue to court. Ford had founded Ford Motor Company in 1903, stating, "I will build a car for the great multitude." The patent he and his allies were fighting would prevent that promise from being fulfilled. A court battle raged from 1903 until 1911, when the US. Circuit Court of Appeals ruled that the patent was valid, but that it covered only two-cycle engines. By this time, Ford cars and most others being manufactured used four-cycle engines. The independents were free to begin mass production of the machine that changed the world.

Although it sounds obvious today, job one for the fledgling automotive industry was to design and build cars that were reliable enough for people to buy them. There were hundreds of innovations that were necessary to ensure reliability, from electric spark plugs to suspension systems, from inflatable tires to electric headlights. Starters, braking systems, steering mechanisms, rubber gaskets for windows, and the window glass itself, all had to be invented and perfected. Manufacturing improvements and business efficiencies combined to reduce the prices of cars to a level that an average income family could afford to purchase one, and the advent of instalment financing made it easier to make the payments. By 1922, Ford could sell a Model T for the very affordable price of $295.

Then there was the infrastructure that had to be developed. New roads were built and old roads paved, sign conventions were established, and rules of the road were agreed. Gasoline (petrol) refineries were built and standards defined for refining the fuel. Service stations were introduced that sold the fuel and other products for the car, and provided both repair and maintenance services.

At the point that all these developments came together—price, dependability, infrastructure—people could begin buying cars for everyday use, and competition with other forms of transportation could begin. Once speed limits on motor-driven vehicles were lifted, the increased range of this mode of transport versus our own two legs or the animal driven variety gave the automobile and irresistible advantage for both private and commercial use. Instead of being limited to an average of 2.7 miles per hour on foot, or 9.5 miles per hour on bicycle, we could move at average speeds of 25 miles per hour on the same roads as we used previously, and even faster on newly-built special highways.

Figure 3: Automobile Developments 1900 - 1951

Year	Significant Events in Automobile History The First 50 Years
1900	First steering wheel instead of a tiller
	First driver's license issued, New York City
1902	American Automobile Association founded
1905	First cars sold on installment plan
1907	First car with left-hand steering in U.S. – Ford
	General Motors formed with Buick, Cadillac, Oakland (Pontiac) and Oldsmobile
1908	First Model T Ford
1911	First Electric starter – Cadillac
1913	First moving assembly line – Ford – producing 1000 cars per day
1914	First stop sign – Detroit
1916	National Highway System established with President Woodrow Wilson signing the Federal Aid Road Act
1918	First three-color traffic signal – Detroit
1924	One in seven Americans owns a car
1925	U.S. government takes over the numbering of federal roads.
	Octane scale introduced
	Safety glass introduced
1928	Coast-to-coast bus service introduced
1930	First cars wired for radio
1936	Porsche builds the first Volkswagen
	First production diesel cars – Mercedes
1938	Buick introduces first electric turn signals
1940	Sealed beam headlights
	Packard introduces first air conditioning system
1942	Civilian car production halts for WWII
1945	Passenger car production resumes – July
1946	First power windows
1951	Power steering commercially available – Chrysler

Now here is where humankind made a crucial evolutionary decision. We could have chosen to stay where we were, to keep our jobs and homes and shopping and schools and recreation facilities all close to one another, and

23

we could have benefited from the added speed provided by the automobile to have gotten to where we needed to go much faster than on foot or with the help of beasts of burden. We could have taken the time we saved getting to where we needed to go and used it to spend more time in one place or the other, like sleeping later, or getting home to our families earlier. But we did not choose this option. Instead, we chose to move all the places around so that they were all farther apart. We then ended up spending the same amount of time (at least initially, before congestion) going from one place to another as we did before we had higher speed transportation.

The first thing many people moved was where they lived. They moved out of their tight city quarters and joined the small plot farmers in the suburbs. The word **suburb** is Latin in origin, *suburbium*. A *suburb* is a district lying immediately outside a city or town, especially a residential section outside of the city boundaries, but adjoining them.[28] The word has become synonymous with "urban sprawl", but it did not always have negative connotations. In the residential suburbs of the first half of the 20th century surrounding New York, Boston, Chicago and Los Angeles (yes, Los Angeles) in the U.S., and the great metropolises in Europe, those who could afford to move out of the cities and make the daily commute by train or streetcar, or those who found work in these newly founded settlements, lived in communities that contained the structure and pattern of urban settings. Note the following description of the Los Angeles suburbs:

"By the 1920s, an extraordinary 2,500 mile inter-urban train system called the Pacific Electric Rail System, or Red Car, was virtually complete, allowing people from all over the region to commute to its center. The intense growth of train suburbs and charter towns surrounding Los Angeles and the idea of a dense downtown employment district developed simultaneously. Extraordinary places such as Glendale, Burbank, Beverly Hills, San Marino, and many others were founded, planned and built as isolated, self-sufficient towns with a balance of civic, commercial, recreational, and residential uses. Along with their equivalent neighborhoods within Los Angeles city limits, they offered a small-town atmosphere and lifestyle to their residents away from the congestion of the regional employment center Downtown."[29]

The process of large-scale suburbanization began in post-World War II United States. The country's industry had not been ravaged by war, as had the industries of most countries in Europe and Asia. Returning soldiers, sailors, marines and airmen married, started families, and began working in the offices and factories that were converted back from military to commercial activities. New residential and living patterns developed to accommodate this swell of new families and surge of children.

Single-use zoning prevailed during this period. It was generally thought to provide healthier and more harmonious environments, in spite of warnings by some sociologists and city planners at the time that single-use zoning created both sterile and potentially dangerous environments.[30] The single-family house on a quarter-acre lot placed on a cul-de-sac (a.k.a. dead end street)[31] became the ideal. With densities reduced from 72,000 people per square mile (26,000 per square kilometer) in the places these new suburbanites had evacuated, such as Manhattan, to less than 2,500 people per square mile (900 per square kilometer) in the suburbs, neighborhood facilities that were reachable on foot, such as schools, shops or even churches, were no longer practical. Catchment areas that had contained close to 20,000 people within a five-to-ten-minute walk—a comfortable maximum for carrying grocery bags—or up 150,000 people within a fifteen-to-thirty minute walk for secondary school children, now contained too few people to make these local facilities viable.

The stage was now being set for ushering in the era of the automobile. An essential piece of this puzzle was getting food and other essentials for daily living from the store to the home.

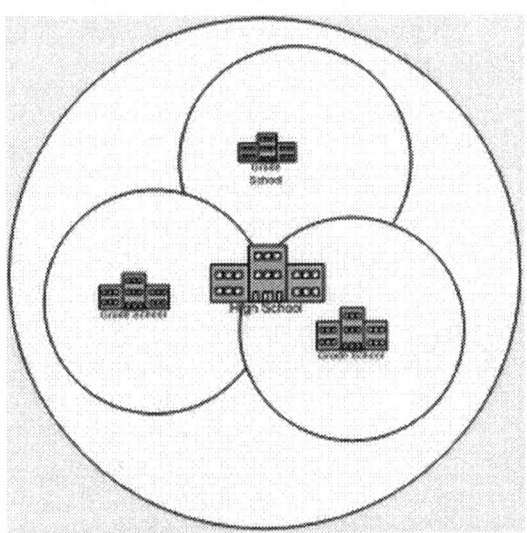

In high-density urban regions, children can walk to grade schools and high schools since the catchment areas are relatively small but contain a large number of families. In low-density suburban regions, the catchment area for a grade school may be the size of an entire community, and that for a high school the size of several communities.

The supermarket was the cornerstone of centralized shopping. **Supermarkets** began to appear gradually in the U.S. in the early part of the 20th century, and their development is completely intertwined with the suburbanization of residential areas and the proliferation of the automobile. The supermarket was "invented" by Clarence Saunders with his *Piggly Wiggly* stores.[32] He opened his first Piggly Wiggly in 1916 in Memphis, Tennessee. By 1922, there were 1,200 Piggly Wiggly stores in 22 states. By 1932, the number of stores had grown to 2,660. *The Great Atlantic and Pacific Tea Company*, or *A&P*, founded in 1859, was one of the most successful early national supermarket chains from the early 1920s up until today.

Before supermarkets, shopkeepers fetched and often wrapped the products requested by the customer, measuring or weighing out the amount desired. Except for the small, general stores, shops were specialized, and shop-keepers had distinct skills, such as butchering, baking, dairying, or fish mongering to name just a few. Mr. Saunders came up with the novel idea of letting customers choose their own products which had been portioned, wrapped or boxed and placed on shelves arranged along isles. Specialists were assembled under one roof so that instead of having to visit multiple stores, the shopper could buy a week's worth of groceries and household needs in one place at one time. Eventually, the shopping cart was added for convenience. When the cart was full, or the shopping list exhausted, the customer came to the check-out counter to have the selections registered and bagged.

The early supermarkets were a storefront on a main street, and were located in the central area or neighborhood sites, just like the modern supermarkets in older U.S. cities, like New York and Boston, and European cities like London and Paris. If there was parking, it was in the rear of the facility or in a lot to the side of the building, but it was minimal. When suburbanization began, supermarkets became the anchor in many of the early shopping centers. The design paradigm became large, single-storey structures set far back from the road with large, paved parking lots in front and docking facilities in the back. Early supermarkets were located in the cities because that is where the people were. As people moved to the suburbs, the supermarkets and associated stores followed.

The vicious circle had been started. The economic viability of small-scale local facilities, whether they were shops or schools, was damaged by the lack of a customer base of sufficient size to support them. Since there were no local facilities, people were forced to drive to large supermarkets or shopping center department stores for even their most immediate needs. The same was true for putting school buses on the roads. Once they were there, it made more sense to keep them there to pick up as many children as possible and bring them to a large, centralized facility, than to build smaller facilities with fewer children.

If you cannot walk anywhere anyway, why spend money on building sidewalks? They just take space from the road, and from the homeowner's lawn. If there aren't enough children within a reasonable walking or bicycling distance to use playgrounds, why build them either? Parents can drive their children to large play facilities, or the children can entertain themselves at home in their own swimming pool or play gym set.

Another new invention that appeared at the same time that America was being suburbanized was children's television.[33] It started with Saturday morning shows, but then spread to all days and times. With fewer options for children out-of-doors, watching TV became a natural alternative. Many parents viewed it as a God-send, a combination baby-sitter, entertainer and teacher (e.g. Sesame Street). Is television part of the reason there is traffic congestion? Indirectly, yes. It helped to make life bearable for parents who had to find ways of keeping their children entertained inside the confined boundaries of their suburban estates in between the times they were chauffeuring them to activities all over the region.

No single person or organization decided that America was going to undergo such a complete transformation, from a nation of cities and farms to a nation of sprawl. It just seemed to happen. And it happened very quickly. In less than twenty years, from 1960 to 1980, the country was a totally different place. Words like *megalopolis*[34] and *conurbation* came into usage to describe the phenomenon. These words refer to an aggregation or continuous network of urban communities which have, through population growth and expansion, physically merged to form one continuous built-up area. The process started somewhat later in Europe, but the effect was the same.

There were strong commercial interests pushing these changes. The cities had been built out by the middle of the 20th century, and the cost of land was at a premium. High land prices in the largest cities dictated vertical building solutions. Floating slab foundations, steel frame building construction techniques, the elevator, and above all, Elisha Graves Otis's invention to stop the free fall of the elevator, were developments of the late 19th century that allowed the skyscrapers of the early 20th century to be built. But there was a limit to both the height and the space to erect these towering structures.

Employee parking spaces came with a very high premium. As a larger percentage of the workforce commuted by car each day, the prices of public parking rose in parallel with demand. All-day, worker-parkers competed with shopper-parkers who drove into the city to visit the major department stores and specialty shops, and higher prices began to discourage these shoppers. Parking was free in the suburban shopping centers, and the concept of the *mall* was beginning to attract more people to both shop and promenade. The shopping mall is a variation on the middle eastern bazaar, the London Arcade, the Italian galleria, and the Viennese pedestrian-only precinct. It is a collection of shops that are accessible from the pedestrian area. The pedestrian areas and shops can be enclosed, and even multi-storey, so that access from street or walkway to shops can be free flowing. Or the pedestrian streets can be open to the air, like they are in many European cities, and the shops accessed through normal doors. Compared to shopping in cities, which in the U.S. began to take on a tattered look in the late 1960s, shopping malls were new and offered a more enjoyable experience.

Congestion and pollution in urban areas increased as the numbers of cars moving about the streets multiplied. In cities such as London, where buses, lorries and taxis were powered by first generation diesel engines, air quality was definitely at the unhealthy level. Smog alerts started to be broadcast in Los Angeles, New York, Athens and Denver. Crime increased in the cities for a combination of complex reasons, but in part as a result of an influx of poor families who were not used to urban living taking over the apartments of those who had fled to the suburbs. Racial tensions and riots in the late 1960s accelerated the flight of inhabitants and businesses from cities.

The logic of building new places of work in the suburbs, to where the workforce was moving, was inescapable. One example is Digital Equipment Corporation (DEC). It was the second largest computer company worldwide in 1987 and reached the *Fortune 50* in that year. The company hit hard times in the late 1990s and was eventually acquired by Compaq Computers in 1998. Ken Olsen founded DEC in 1957. He had been working at the MIT Lincoln Labs outside Boston at the time he founded his new company. He could have decided to place his start-up in an abandoned dockside warehouse in Boston, or even in Chelsea, East Boston or Roxbury. He chose an abandoned mill in Maynard, Massachusetts instead. Back in 1957, Maynard was in the middle of nowhere, about thirty miles (50 km) from central Boston. But it happened to be not that far from where Mr. Olsen and his family lived in the western suburbs of Boston. Mr. Olsen surely knew that he could get all of the low-cost labor he needed to build his computers from the small towns in the region, like Maynard, Hudson and Acton. These towns once had thriving factories making shoes and leather goods, furniture, fabric and clothing. Most of the factories stood idle, like The Mill in Maynard, and unemployment was very high. Mr. Olsen probably also understood that he could more easily convince his bright, young engineers to move out with their families to the fields and farms of the western suburbs, than to suggest that they struggle with the commute into Boston each day, or worse, to recommend that they live in the city.

Hundreds of new computer start-ups followed Digital's lead and located their fledgling companies in towns around the outer edges of Boston's suburbs. Prime Computer started in Framingham; Wang built in Lowell; Data General located in Westborough; Computervision settled in Bedford; Applicon and Xyvision were in Burlington; and Digital expanded its facilities in all directions from its base in Maynard.[35]

While new companies chose suburban locations for their initial locations, many established companies began leaving the cities for these same locations in the late 1960s. By the early 1970s, they were fleeing in droves.[36] According to the Time Magazine Article, from 1967 through 1971, twenty-two large companies moved their headquarters out of New York City, eleven more had plans to do so, and another few dozen had the issue up for discussion.

Why, suddenly, did all of these companies have the collective urge to move? Economic considerations are always cited as the primary reason for relocating a company's headquarters, but, in the end, it is the CEO and the board of directors who decide when a company's stakes will be picked up, and to where the company will be re-located. It is invariably within a comfortable commute of the CEO's place of residence, whether that place is Lincoln, Massachusetts or Scarsdale, New York, Southampton, England, or Versailles, France.

Clearing the way and making it acceptable to work in a suburb rather than a city was a change in the definition of personal success between the heydays of cities in the U.S. in the late 19th century. Back then, successful men such as Elisha Graves Otis, Potter Palmer, Philip Armour, George Pullman, William Astor and Cornelius "Commodore" Vanderbilt walked to work from their Chicago and New York City mansions. The next generation of wealthy businessmen rode in special cars from their homes in Winetka, Illinois and Pleasantville, New York. Still later, they made the journey from home to work in chauffer-driven limousines. The Chicago magnates fled the stockyard stench of the city for the fresh air of the suburbs when transportation made it possible, and when the social stigma of living in the suburbs was removed. When living in cities or comfortably commuting from the country was no longer an option, it was time to move the business.

At the point when suburbs were mature enough to provide all of the facilities needed to run successful businesses, it was possible to make these moves. Memberships in the downtown businessmen's clubs were traded in for membership in the expensive and exclusive country clubs where the entire family could take advantage of the facilities.

In 1970, 25% of the office stock in the U.S. was in suburban locations. By the early 1990s, the amount had risen to 57%. Because it is in offices that Americans now work (40% of all jobs were in offices in 2001), there are more jobs in the suburbs than in the cities. In fact, it is no longer relevant to talk about city and suburb, with jobs located in the core and residences located in concentric rings surrounding the core. And it is not only the new metro areas of the west and southwest where this phenomenon is occurring. Milwaukee, an old urban center in the Midwestern part of the U.S., lost 14,000 jobs in its central business district between 1979 and 1994, while the inner ring suburbs gained 4,800 jobs, and jobs in the outlying areas swelled by an astounding 84,000![37] People now live, work, shop, recreate and congregate, in what used to be the suburbs, not needing or desiring to visit what was their city center.

The problem from a traffic point of view is that we usually do not do all of these things in the same *geographically contiguous area*. In the U.S. today, people live in one area, work in a second, often send their children to schools in other areas, and shop and recreate in several different locations, some near and some far away. This cross commuting is what makes collective transportation an impossibility. As long as people lived, recreated, and shopped in their local community, and worked "downtown", the spoke-and-hub train commuter train solution was viable. In the cities of sufficient density, bus networks could move people between residential areas and industrial or office areas with relative efficiency, or, in the older suburbs, between their residences and the local civic, cultural and commercial centers where the train station was located and then to downtown. Initially, increased car use inside the cities caused a deterioration in bus service, and eventually, the dispersion of both people and jobs all over the region destroyed the operational rationale of all collective transportation.

Spreading households out over a larger area decreases the density. Unless the numbers of bus lines are increased, it means that potential bus riders must walk longer distances to a bus stop, which is inconvenient for some, and impossible for others, like the elderly and infirm. To get the same number of riders on their buses, routes have to be fewer covering larger catchment areas. These two forces work against each other, forcing individuals to seek other alternatives. The private car is the easiest to hand. Taxis are a second option for those who cannot drive or do not own a car. When bus lines stop functioning as feeders, fixed rail options also lose ridership. Even suburban park-and-ride schemes lose some attraction if potential riders must back track significant distances to reach stations.

The situation is not yet as bad in Europe as in the U.S., as the table below shows. There are 8,760 hours in a non-leap year. The average American spends approximately 6% of that time in a car (541 hours / 8760 hours), either as a passenger or as a driver. Work accounts for around 20% and sleep another 30%. That leaves around 44% of time to do whatever else one does in life to make it meaningful. In Europe, where an average vacation period is five-to-six weeks, instead of the measly two in the U.S., work accounts for only 17% of time, and driving or riding only 3% (274 hours / 8760 hours). The average European therefore gets to spend an extra 6% of his or her time sleeping or enjoying fruits of their labors.

Figure 4: Average Time in Vehicles: US vs. Europe

	USA	Europe
Total hours per year	541	274
As Driver	340	183
As Passenger	201	91

Source: Roland Berger analysis (2000)

Europeans should not be too smug. As anyone who has been locked in traffic in and around Paris, or anywhere on The Netherland's motorway system, or on an Autobahn in Bavaria heading south on a Friday evening in summer, or just about anywhere in the vicinity of a city in the U.K. can attest, traffic congestion is real and getting worse.

If we are going to eliminate traffic congestion, we are going to have to do one of two things: either force everyone to stay at home, or rebuild our urban regions to work with collective transportation. The first option is totally impractical. This would mean returning to the cottage industry of the pre-industrial era. It may sound idyllic to many who are truly worn out by the daily traffic grind. It was an answer that was tried by society drop-outs in the 1970s, but the ones without personal fortunes did not last very long in the log cabins in the mountain. So we are left with the second option, rebuilding. That may appear to be not only impractical, but next to impossible. Evidence to the contrary comes from a place that most might think totally unlikely: Los Angeles.

"After fifty years of doubting, it has finally become clear that freeways don't relieve congestion but induce it. The promise of free and rapid mobility by car through the LA basin has been dashed. Commuting across the region is out of control. The half-hour freeway ride between Pasadena and Santa Monica is now a distressing one and a half hours in the morning and evening. As a result, alternative means of mobility have emerged and prospered. In the last decade, rapid bus and rail transit systems have become regional in their reach. Working at home, or closer to home is increasingly seen as a preferred way of living in Los Angeles. Transit-oriented development at the scale of whole neighborhoods is becoming an alternative to both extensive car ownership and time-wasteful commuting by car."[38]

The New Urbanism movement among architects and planners all over the world is not based on political utopianism and shrouded in anti-growth rhetoric. More importantly, it is not based on the promotion of an architectural style, such as neo-classicism in 1890-1920, modernism in 1920-1960 or post-modernism in 1980-1990, which proposed new city forms created from massive application of an architectural vernacular. New Urbanism is anchored in the view that the principles that make neighborhoods livable, such as placing shopping, recreation and school facilities within walking and cycling distance of one's home, and fostering office and commercial development such that they support the use of collective transportation, make entire regions, and eventually whole countries, more livable. They argue convincingly that these principles can be extended to entire regions to create environmentally sustainable habitats that do not force everyone to use an automobile to execute the most basic daily functions. It is in this kind of thinking that the long-term solution to traffic congestion lies.

And, finally, when discussing the roots of traffic congestion, we cannot leave out mention of *The Kingston Trio*. The lyrics of their song,

"Charlie on the MTA", helped to put the "babyboom" generation off public transportation and into cars to avoid the fate of Charlie, the hapless rider of Boston's subway system.

> *Let me tell you the story*
> *Of a man named Charlie*
> *On a tragic and fateful day*
> *He put ten cents in his pocket,*
> *Kissed his wife and family*
> *Went to ride on the MTA*
>
> *Charlie handed in his dime*
> *At the Kendall Square Station*
> *And he changed for Jamaica Plain*
> *When he got there the conductor told him,*
> *"One more nickel."*
> *Charlie could not get off that train.*
>
> *Chorus:*
> > *Did he ever return,*
> > *No he never returned*
> > *And his fate is still unlearn'd*
> > *He may ride forever*
> > *'neath the streets of Boston*
> > *He's the man who never returned.*[39]

Figure 5: Automobile Developments 1952 - 2006

Year	Significant Events in Automobile History The Second 50 Years
1952	Four barrel carburetors – Buick, Oldsmobile, Cadillac
1953	Chevrolet Corvette has first fiberglass body
	First 12-volt electrical systems
1955	Pushbutton automatic transmission - Chrysler
1956	First electric door locks
1957	Cruise control introduced
1958	Government mandates retail price stickers on cars
1957	First electric trunk locks
1959	First U.S. built compact cars – Chevrolet Corvair, Ford Falcon, Plymouth Valiant
1963	Front seat belts offered as standard
1966	First electric fuel injection system invented in Britain
1973	The 1973, Oldsmobile Toronado was the first car with a passenger air bag intended for sale to the public
1974	Psychologist John Voevodsky invents third brake light
1978	Tennessee in the U.S. becomes the first state to require child safety seat use.
1980s	Just in time delivery introduced in auto manufacuring
1985	The Lincoln Continental (Ford Motor Company) is first American car to offer antilock braking system (ABS), which is made by Teves of Germany.
1992	U.S. Energy Policy Act of 1992 encourages fuel efficient vehicles.
1996	General Motors introduces OnStar in Cadillac.
1997	Toyota introduces the Prius Hybrid-electric vehicle (HEV)
1998	Crude oil prices near $11 per barrel.
2001	Sales of SUVs, pickups and mini-vans top sales of cars in U.S. for first time.
2006	Crude oil prices top $75 per barrel, and sales of SUVs, pickups and minivans drop over 25%.

Too Much of a Good Thing

WHAT IS TRAFFIC CONGESTION?[40] I would argue that it is too much of a good thing. Cars provide an unbeatable form of transportation: flexibility; comfort; convenience, and cost-effectiveness. Trucks cannot be equaled for moving goods in the manufacturing and retailing world that exists today. New motor vehicles use significantly less fuel and pollute the environment measurably less than their predecessors. The main problem is that there are just too many of these great inventions trying to use all the roads we have built for them—seemingly all at the same time.

In 2003, there were 204 million vehicles in the U.S. and 191 million licensed drivers. Looking at it another way, there were 107 million households, each with an average of 1.9 cars, trucks or sport utility vehicles, and there were 1.8 drivers in those households. In other words, there are more vehicles registered in the U.S. than there are licensed drivers. This appears to be too much of a good thing, and the result is congestion.

Between 1970 and 1996, the distance that people in the U.S. drove annually grew four times faster than the population, twice as fast as licensed drivers and eighteen times faster than new roads were built.[41]

Figure 6: Growth in U.S. automobile usage 1970 - 1996

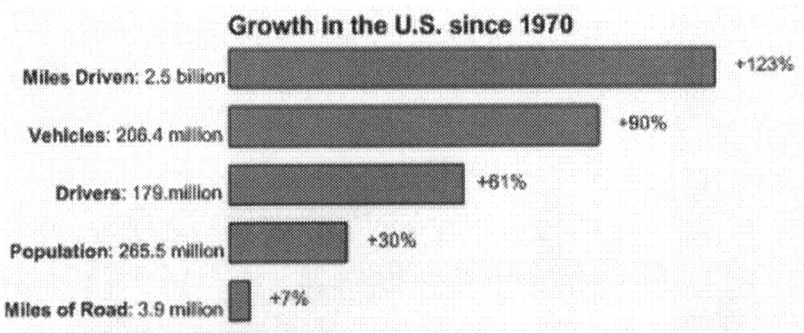

Source: U.S. Federal Highway Administration (1998)

The causes of congestion have been catalogued by government agencies around the world in countless reports and documents. They all say about the same thing, that approximately one-half of the causes are *predictable*, and the other half are *inevitable*. Something is *predictable* if it is possible to tell beforehand what will happen. Something is *inevitable* if it cannot be avoided. For example, scientists have predicted with a fairly high degree of certainty that our solar system will collapse in approximately 4.5 billion years. Our sun will do what many suns before it have done: It will run out of energy and explode. When that occurs, it is inevitable that all things living on the earth will be destroyed. (By the way, this is a pretty good reason for an active space exploration program, one that can get all the folks who may still be around in 4.5 billion years over to a new solar system.) A more mundane example of a predictable event is a traffic jam at a heavily used intersection on any workday during the morning and evening rush hour. Although we cannot predict when, inevitably, on some days, one driver at that intersection will rear-end another driver by braking too late or accelerating too early, and long travel delays will ensue.

Accidents involving automobiles are inevitable, since neither the humans who drive them, nor the automobiles themselves, are capable of perfect performance. Once an accident has happened, it is inevitable that there will be some form of traffic disturbance, the degree of the disturbance being dependent on the number of other vehicles traveling on the same stretch of road behind the location of the incident. Most of us have passed by a car stopped by the side of the road, the driver obviously out of fuel, waiting for roadside assistance to arrive. If this occurs during the workday rush hour, or on a Sunday evening when everyone is coming home from their weekend in the country or at the shore, it is certain to cause a traffic jam. If it happens in the middle of the day, it has little effect on traffic flow.

Figure 7: Reasons for Traffic Congestion

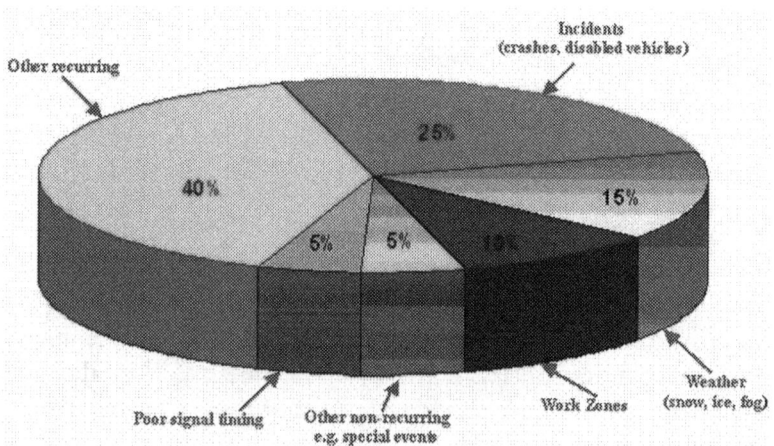

Source: U.S. Federal Highway Administration, Office of Operations (2000)

The diagram above is from the U.S. Federal Highway Administration.[42] The *predictable* events in the pie chart, fully 60% of them, are "Work Zones" (10%), "Poor signal timing" (5%), "Other recurring events" (40%), and "Other non-recurring" events (5%). The "Other recurring events" are the most numerous and include daily jam-ups at highway on- and off-ramps, roundabouts (advanced versions of traffic circles)[43], entrances to schools, businesses or commercial facilities. These tie-ups usually disappear totally after the morning rush hour, re-appear to a lesser degree during the lunch hour, and then come back in force during the late afternoon when people begin leaving the facilities to return home.

The *inevitable* events, the remaining 40% of the reasons for traffic congestion, involving "Weather" (15%) and "Incidents" (25%), are those that occur because both people and machines are not perfect, and because we continue to try to go about our business as usual on days when the powers who are in charge of the weather would have us do otherwise. Inevitable events are caused either by people making poor judgments while they drive, or about when to drive, or by the failures of the vehicles they are driving, or the infrastructure on which they are driving (e.g. bridges failing or roads opening up into sink holes).

Let's first look at the *predictable* events. It is not difficult to understand why there is a traffic jam at the same place and time every week day at peak travel times. The office park entrance; the intersection of two major roads; the on- or off-ramp between an expressway and a large employment and shopping center; or the roadway itself: They were not designed to accommodate the number of vehicles using them at peak periods. Didn't the engineers who designed the solution understand the problem, or weren't they capable of developing a workable solution? Didn't the developers allocate enough money to the project, or didn't the communities set the guidelines clearly enough that there should be no traffic congestion?

Ask the engineers and they will say that their design was predicated on a certain number of vehicles, but when it began operating, more people used it. This sounds like *Downs's Law* in operation—road usage always increases until a state of equilibrium is reached, and equilibrium always equals congestion at peak hours. If this is always the case, why not double the estimates in the first place, or why not restrict the number of people who are allowed to use the facilities, thereby limiting the number of people who could potentially cause a traffic jam? Ask the developers, and they will say that they are working to the building code. Ask the community planners and they will say that they are trying to maximize the number of jobs and the tax base. Good intentions; bad results.

This is the crux of the problem with car and truck traffic. There is no internal mechanism to control demand, except the cost of using the vehicle. This is what is meant by stating that congestion is "the controlling mechanism for a scarce resource." It is not that there are not enough roads, or that they have been built in the wrong places. As anyone can see when driving outside of rush hour times, most of the road surfaces are unused. There are large stretches of empty space in between cars, even during rush hour. We are taught in driving school to maintain a safe distance between the front of our own vehicle and the back of the vehicle in front of us. When an inconsiderate driver insists on passing to the right and left in heavy traffic, and pulls into the safe space we had created behind the vehicle we were following, we automatically brake to re-build that space.

"A vehicle's road space requirements increase with speed, because drivers must leave more shy distance between their vehicle and other objects on or beside the roadway. Traffic flow (the number of vehicles that can travel on a road over a particular time period) tends to be maximized at 30-55 mph on highways with no intersections, and at even lower speeds on arterials with signalized intersections."[44]

The problem is that where and when capacity is most needed, it is not available. We have not designed our roads to expand and contract with increased and diminished demand (although reversible lanes is the right step in that direction). We cannot move capacity from one place to another, like we can add leafs to a dining room table and bring in the folding chairs when all the relatives come in for a holiday meal. We have also not yet put vehicles into production that optimize road usage by bunching up cars and trucks to bumper-to-bumper distances while allowing them to travel at reasonable speeds and still avoid collisions.[45] We can't make cars smaller when congestion is heaviest and enlarge them again when there are fewer cars on the road. We have not yet found a way to deliver more road space when and where it is needed on the roads for more cars. Imagine if traffic could move like water, flowing faster and rising higher when constricted to tight spaces, and always knowing how to find the straightest fall line from origin to destination.

We (planners, government officials, society) relaxed the *No Congestion* criteria in favor of other criteria that we judged more important. These other criteria included: building roadways at the lowest possible cost; developing cars that were safer to drive and did a better job of protecting the occupants in case of a crash; and, maximizing the number of workers in the office parks, the number of inhabitants in residential areas, the number of stores in shopping centers, and the number of children in schools.

Now, government officials want to reduce or eliminate the predictable events by adding <u>external</u> mechanisms to control demand, like charging drivers fees for using the roads and their cars. The stated logic of these fee-charging proposals is that by restricting demand there will be a better utilization of the available supply. Ignoring for the moment the inherent unfairness of indiscriminate charging schemes which favor the wealthy—who can afford to pay any size road usage charge—over those who are just scraping by, there is ample evidence that removing one class of drivers from the roads (i.e., those who cannot afford the charges) simply opens up space for another class, whether it is drivers who were formerly taking public transit and who had previously avoided the roads because they were congested, or commercial vehicle drivers who had avoided certain roads because they were constantly congested .

Instead of making it more expensive to drive and thereby punishing drivers for the mistakes that government bodies and business interests have made in the past —which is what we do by raising fuel taxes or imposing toll collection and congestion charging schemes—we need to work on <u>correcting the roots of congestion with long-term fixes</u>, while at the same time, in the short term, doing everything possible to remove the bottlenecks and increase capacity on our roads. What can be done to address the short-term, predictable problems without punishing the victims?

We apologize for the inconvenience while we make your roads safer and better for you in the future.

Road maintenance work is a major cause of traffic congestion. How often have you thought that it would be better to leave the roads in a poor state of repair rather than having to suffer the constant congestion caused by improving them. Better work zone management could go a long way toward reducing both traffic congestion and driver frustration. One suggestion is to tie total payment for the road works contract to the rate of traffic disturbance it causes, encouraging working at night and during off-peak times, and restoring total capacity during peak times. Another suggestion is providing alternate routes around work zones, informing drivers about the routes, and providing staff to ensure that the routes are safe, both for the re-routed drivers and the people who live, work or play along the detours. When possible and when viable alternate routes are available, roads being improved should be completely closed to allow the work to proceed at a faster pace, in greater safety for the workers, and to simply eliminate the problem of congestion on that particular stretch of road.

Another strategy for work zones is to use innovative materials and construction techniques to reduce construction times. Road re-surfacing in Sweden must be done often because of the heavy use of studded tires. Ruts are worn into the roads that need to be filled every few years. Rather than simply filling the grooves, the entire surface is repaved. Innovative machines literally eat up about fifteen centimeters of the surface, process the tar and gravel, and lay it down again at the other end. This work is generally performed during the night, one lane at a time, with a minimum of disturbance to traffic.

Limiting the number of cars that have access to facilities to a number that will not cause traffic jams seems like a logical action. This has been done in Italian cities, where access to most central cities during week days is controlled by the last digit of the vehicle's license plate. Odd numbers are allowed on odd numbered days (e.g. all ones and threes and fives and sevens and nines allowed on the first, third, fifth, etc. of the month) and even numbers allowed on even numbered days. Yes, a wealthy person with two cars can buy license plates with odd and even numbers. However, this solution seems to be working to control traffic and reduce pollution. It has the added effect of increasing the use of collective transportation because there are no incentives for transferring out of buses and trams to cars to fill the space that has been left vacant.

For this solution to work, there have to be alternatives. Simply installing a guard at the entrance of an office park who would turn away cars with the wrong license number-to-day match is not a solution. There needs to be alternative transportation from a parking facility that is not subject to congestion to the office park or school or commercial facility. This alternative transportation must run often, cost no more than driving directly to the facility, and be as safe. Boston, Massachusetts has done a superb job of building multi-modal, park-and-ride facilities at its edges. Its streets are still congested, but they certainly would be worse without its commuter rail, extensive bus network and its subways and trams.

In later chapters I will take up the issues of traffic congestion caused by school buses and parents driving their children to school, and the reduction in road space caused by the increase in the number of large trucks on our highways, arteries and local roads. Getting our children out of vehicles and onto sidewalks and bicycle paths, and separating large commercial trucking from the rest of the transportation network, would go a long way toward reducing traffic congestion.

&

Now let us look at the unforeseen incidents of every kind. Incidents on the roads, such as accidents and car failures, and weather, are the second and third most common causes of congestion. If accidents are a major cause of traffic congestion, reducing the number of accidents should contribute to reducing the amount and severity of traffic congestion. Any disturbance, no matter how small, to a system that is only marginally balanced will immediately cause an imbalance. The road transportation network is a system that, at its best, is only marginally in balance, and, at its worst, is constantly on the verge of total breakdown. Certain stretches of particular roads are always congested. Any incident, from a flat tire to a fender-bending accident to a full-fledged, multi-vehicle pile up, will create havoc that can extend to the horizon in both directions of travel approaching the scene.

Figure 8: Vehicle-related deaths in selected countries

International Road Traffic and Accident Data, 2001				
Deaths per 100,000				
	Inhabitants	Vehicles	Number of Deaths	Millions of Inhabitants	Vehicles (000s)
Great Britain	6.0	12.1	3,580	59.8	29,521
Norway	6.1	10.2	275	4.5	2,699
Sweden	6.2	11.4	554	8.9	4,872
Netherlands	6.2	12.2	993	16.0	8,169
Switzerland	7.6	11.6	544	7.2	4,707
Japan	7-9	12.6	10,060	127.3	79,602
Denmark	8.1	17.7	431	5.3	2,434
Finland	8.4	17.4	433	5.2	2,483
Germany	8.5	13.3	6,977	82.3	52,487
Australia	9.4	14.6	1,824	19.4	12,477
Canada	9.6	16.4	2,972	31.1	18,192
Ireland	10.7	23.2	411	3.8	1,770
Italy	11.1	16.9	6,410	57.8	37,836
New Zealand	11.8	17.3	455	3.9	2,633
Austria	11.9	18.3	958	8.1	5,228
Spain	13.8	22.8	5,517	40.1	24,250
France	13.8	23.5	8,160	59.0	34,781
Belgium	14.4	25.6	1,470	10.2	5,735
United States	14.8	19.0	42,116	284.8	221,230
Luxembourg	17.5	22.8	76	0.4	333
Portugal	17.6	22.5	1,671	9.5	7,427
Greece	20.2	48.9	2,116	10.5	4,323

Source: 2000 OECD/IRTAD Database

Humans are accident-prone, and not just in their cars. More people are killed in their homes than on the roads. The U.K. Department of Trade and Industry estimated that in 1999, household accidents cost their society around £25 billion per year. There were 42,000 casualties from eating and drinking; house slippers—shoes that people wear inside the house rather than outside—caused 37,000 injuries; flower pots injured a further 3,500. By comparison, in the same year, there were 3,421 vehicle-related deaths in the U.K., and 235,000 people injured as a result of vehicles hitting each other, stationary objects, or pedestrians .[46]

The World Health Organization statistics indicate that there are approximately 1.2 million deaths and 50 million injuries each year in accidents involving motor vehicles. These worldwide figures are expected to increase by 60% by 2020 because of increases in automobile usage. However, the projections are for decreases of 30% in the OECD[47] countries, due to increased safety measures and improved cars, and increases of 80% in the emerging countries, where cars tend to be older and roads less safe.[48]

Why do vehicle accidents occur? There are many reasons. Often there are multiple causes contributing to a single accident. The reasons vehicle accidents occur include the following[49]:

- Driver distraction – This includes tuning the radio or air conditioning or other devices in the vehicle, talking with passengers, eating or drinking, putting on makeup or performing other grooming activities, dealing with children or pets in the back seat, picking up dropped items, or any activity that takes the driver's eyes from the road for a critical number of seconds. If you have read John Irving's classic novel, *The World According to Garp*, you know that there are other forms of distraction, even in a non-moving vehicle, that can lead to disastrous results.

- Driver impairment – This can be the result of tiredness, illness, alcohol, drugs (both legal and illegal), or a physical limitation that manifests itself under certain conditions, such as reduced night vision.

- Vehicle or driver deficiencies - Inability of the driver to operate the car with adequate control. Accidents in this category might be due to inadequate driver training, or poor design of the car or its controls.

- Mechanical failure – This includes such problems as flat tires or tires blowing out, brake failure, axle failure, steering mechanism failure, among others.

- Dangerous road conditions – This includes ice, snow, water or other substances making the road surface slippery; or obstacles that have fallen onto the road, or damage to the road surface, that make it difficult to control the vehicle..

- Excessive speed – This includes speeds that exceed those for which the road and/or conditions allow for safe operation of different types of vehicles.

- Road design and layout – There are stretches of roads that are well known for being hazardous or "death traps".   Reasons include alignment, visibility, camber and surface conditions, road markings, among others.

The worst time for fatal accidents in the U.S. is between midnight and 3 a.m. on Saturdays and Sundays. Most of the fatal crashes (57%) involve only one vehicle, compared to 30% injury crashes and 31% property-damage-only crashes. Over one-half of the fatal crashes occurred on roads with posted speed limits of 55 miles per hour or more, while only 25% of property-damage-only crashes occurred on these roads. Collision with another moving motor vehicle was the most common reason for all types of accidents, fatal injury and property-damage-only. Collision with fixed objects (e.g. trees, light poles, sign posts, etc.) and non-collisions (e.g. roll-overs) accounted for only 19% of all crashes, but 44% of all deaths. One of the saddest figures is that 40% of all fatal crashes involved alcohol, and for fatal crashes occurring from midnight to 3 a.m., 77% involved alcohol![50]

Vehicle-related accidents and deaths can be reduced. In 2003, 6.3 million Americans were involved in vehicular accidents, and 42,643 died as a result. In 1980, there were 51,000 deaths from vehicular accidents. In the UK in 1975 there were 246,000 vehicular accidents and 6,366 resulting deaths, while in 2003 the figures were 214,000 and 3,508 respectively. In France, from 2002 to 2005, the number of people killed on the roads dropped dramatically, from 8,000 to 5,200. The reduction in deaths in the U.S, U.K., France and other western countries is attributed primarily to safer cars and more stringent control of drunk driving.

Figure 9: U.S. Driving Fatalities: 1994 and 2003

Statistic	1994	2003
Total Fatalities	40,716	42,643
Vehicle Miles Traveled (billions)	2,358	2880
Fatalities per 100 million miles traveled	1.73	1.48

Source: U.S. National Highway Traffic Safety Administration

Sweden, a country for which there are excellent historical traffic statistics, and one of the most stringent public education and enforcement programs for driving while intoxicated, reported in 2005 that 450 people were killed in traffic-related accidents. This is the lowest number of deaths on the roads since 1945, when the country had less than 7 million people and 50,000 cars compared with 9 million people and 4.1 million passenger cars in 2005. Commenting on the low fatality figures, the country's *Road and Transport Research Institute* said:

"If there were as few cars on the roads in 2005 as in 1945, only 9 persons would have died in traffic accidents this year. On the other hand, if cars and trucks were as dangerous today as they were back then, the total number of deaths this year would have reached 15,000!"

The *Institute* credits the seat belt as the most important factor in decreasing deaths. Swedish inventor, Nils Bohlin invented the three-point seat belt. Nils Bohlin's lap-and-shoulder belt was introduced by Volvo in 1959.[51] The use of seat belts in Sweden is obligatory, and has been since 1975. Within a few years of passing this law, the number of deaths on the roads was halved! Today, 90% of drivers in the country use seat belts. The *Institute* reckons that if no one used seat belts, 800 more people per year would be killed in accidents. Further, for every 1% increase in usage, nine more lives would be saved per year.

The single most important road design change in Sweden was the introduction of 2+1 alternating lanes with cable dividers between the opposite directions of traffic. In 2000, 140 people died in head-on collisions. In 2004 that number was reduced to 60 thanks to an intensive program of converting single-lane roads into 2+1 roads.

Another important change for the better in Sweden was in 1967 when speed limits were introduced. Prior to 1967, there were no speed limits outside built-up areas. The peak for traffic deaths was in 1966 when 1,313 died. After speed limits were introduced, deaths declined rapidly and stabilized around 600 per year in the mid-1990s. Also in 1967, the country moved from driving on the left side of the road, like in Great Britain and Japan, to driving on the right side, like in the U.S. and the rest of Europe. This factor is also credited with the reduction in accidents and related deaths in Sweden. It is not the side of the road that mattered, it seems, but the side of the car on which the steering wheel was located. In Britain, the steering wheel is on the right, while in pre- and post-1967 Sweden, steering wheels were on the left. So drivers in pre-1967 Sweden were on the curb side of their cars, away from on-coming traffic. Britain, which continues with left-hand driving, has one of the lowest accident rates in the world.

The next area of focus for Sweden is speed limit enforcement. The road authorities figure that by forcing all drivers to obey the legal speed limits, an additional 100 lives could be saved per year. They are installing hundreds of new traffic cameras on the most accident-prone roads in order to reduce speed limits to legal levels. Between 1999 and 2002, the Swedish Road Administration conducted extensive tests of driver acceptance of in-vehicle systems that either informed drivers of legal speed limits, signaled that they were exceeding the legal speed limits, or actually inhibited the vehicles from driving faster than the legal limits. The aim of the project was to learn more about drivers' acceptance of the different types of systems, to study the systems' effectiveness, and to add to the body of knowledge on the practicality of putting speed limit intelligence into the vehicle.

Figure 10: Number of registered cars in Sweden – 1920-2004

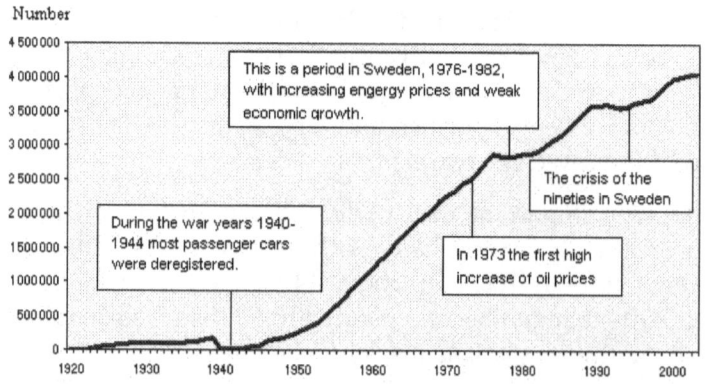

Five thousand vehicles were fitted with the systems, and ten thousand drivers took part in the tests. Driver acceptance was higher than originally expected, and most test drivers felt that the systems should be mandatory in urban areas. The tests also proved that better road safety was possible without significantly increasing travel times. Tests were performed in other European countries with similar results, and there is now a dialog between the public road authorities and the automotive industry to install intelligent speed advisory systems in their vehicles.

Cars have become much safer during the past sixty years with the addition of *passive* safety features, such as seat belts, air bags, crumple zones that absorb the shock of a collision, collapsible steering columns, padded interior surfaces, anti-lock braking systems (ABS), electronic stability programs (ESP), and many, many more improvements. Roads have become safer, with more roundabouts replacing four-way crossings, more traffic calming designs and raised pedestrian crossing zones.

What else can be done to reduce congestion-causing incidents? A good way to reduce the number of incidents is to ensure that the cars on the road are not traffic jams waiting to happen. Keep your car in good repair, have it serviced at recommended intervals, and, at a minimum, try to keep the tank filled with fuel and the cooling system filled with fluid. Keep a careful eye on your tires, and replace them when they are worn out. If the grooves in the thread are less than 1/8" (3.2 mm), it's time to spring for a new set. If you live in a climate where it snows, change to snow tires before the end of November, put them on all four wheels, and keep them on until March.

Do not buy a car in its introductory months. Wait at least one year so that the manufacturer can work out all of the mechanical and software bugs. Software bugs in cars? The average car today contains around fifty microprocessors. Each microprocessor is a computer, and computers run on operating systems and software. As most people who own or have used a computer know, computers stop running for apparently no reason, and software does not always function as we think it should. Computers in cars are no exception. Cars have an increasing amount of electronic parts controlled by computers for a variety of reasons, including the need for sophisticated engine control systems to meet stringent emissions and fuel economy standards, to reduce the number of expensive mechanical parts, and to control new safety, comfort and convenience features.

Vehicle manufacturers are getting much better at ensuring that the software being loaded into their vehicles has been fully tested. They are cooperating in standards efforts that will minimize the risk of incompatibility between the different systems. One day, probably much sooner than in the home and office computer market, car computers will be fail safe. Hopefully, that day will come soon, but it is not here today.

Authorities can help to ease congestion from incidents by being quicker on the scene when accidents and breakdowns occur. In Great Britain, an Incident Support Unit, consisting of 13 vehicles, was put into service around the Birmingham region in 2001 to help clear accidents more quickly and to get traffic moving safely again following accidents. The results of this trial effort were reportedly very good. During the summer of 2005, a similar unit started patrolling the M25, London's ring road, with responsibility for clearing debris, helping police clean up after accidents, moving abandoned vehicles and temporarily closing roads. These so-called *Jam Buster Patrols* were extended to other parts of Great Britain, and by mid-2006, the network was complete. Unlike the police, who perform such duties in most places in the world along with their crime-fighting and traffic violation control, such special-purpose teams have only one set of tasks. They are trained and equipped to perform these tasks, and they can help to make a major contribution to the smooth operation of traffic.

Another good way to quickly get help to the scene of an accident is to have a vehicle equipped with a telematics system, like *OnStar* for GM cars, *Volvo On Call* for Volvo cars and *BMW Assist* for BMW cars. The term *telematics* was originally said to be a combination of telecommunications and informatics, the latter word referring to "the making of information". Literally, **tele·matics** as derived from the Greek root *tele*, meaning "far off" or "distant", and the Roman root *matic*, meaning "to make happen", means "to make something happen at a distance". This is exactly what a telematics system does.

Figure 11: Data and voice flow in the Volvo On Call telematics system

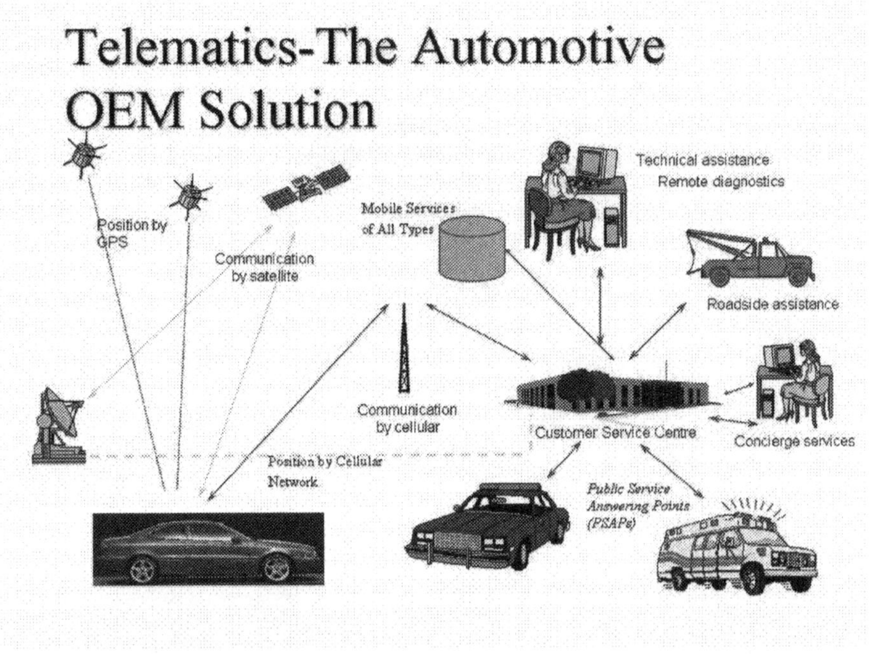

Source: Volvo Car Company and WirelessCar Sweden AB (2006)

With a telematics system, the car automatically sends a message to a specially-equipped customer service center when an air bag is released or a crash sensor is activated. The message includes information about the vehicle and the exact location of where the accident occurred. The location is obtained from the on-board GPS[52] device. Other information is sent as well in the data message, like the temperature inside the vehicle (Is it on fire?); the fuel level (Is the gas tank leaking?); the number of people in the vehicle and much more. In addition to the data being sent, voice contact is established. Even if the occupants are unable to speak, assistance can be directed to the scene.

Telematics extends the passive safety features being built into cars today. The next stage involves building features into cars that actually help to prevent accidents from happening, so-called active safety features. These systems are called Advanced Driver Assistance Systems (ADAS). They support the driver in driving safely, comfortably and economically. Typical ADAS applications already on the market are Adaptive Cruise Control (ACC) and Adaptive Light Control (ALC). ADAS currently perform their functions on the basis of information generated by radar and camera sensors.

Figure 12: Advanced Driver Assistance Systems with map data support
The ADAS Interface Concept

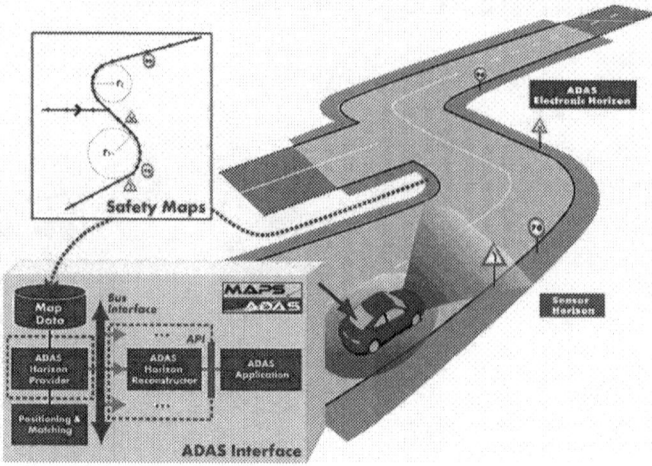

Source: The ADASIS Forum (2006)

The possibility to predict the road geometry with its related attributes ahead of the vehicle should be of major benefit to ADAS applications and offer new functionalities. In the future, the same map data that is used in the navigation system will be used as an additional sensor to see around corners, over the peaks of hills and through buildings.

Wal-Marting the World

Haᴠᴇ ʏᴏᴜ ꜰᴇʟᴛ ᴅᴜʀɪɴɢ ᴛʜᴇ ᴘᴀꜱᴛ ꜱᴇᴠᴇʀᴀʟ ʏᴇᴀʀꜱ that you are sharing the road with more and more heavy commercial vehicles? You are. The number of commercial trucks[53] in operation in the U.S. has increased from around 23 million vehicles in 1974 to 95 million in 2003, a 400% increase. The number of commercial trucks sold in the U.S. per year in these same years increased from 2.7 million in 1974 to 9.4 million in 2003, a 350% increase. This is while the total length of roadways increased by only 5%, from 3.8 million miles to just under 4 million miles. In the United Kingdom and Germany, the increases in the number of commercial trucks in operation and sold were both a more modest 100% over the same thirty-year period.

The U.S. Bureau of Transportation Statistics reported that the fleet of medium and heavy trucks grew 18% between 1992 and 1997. The number of trucks in one of the heaviest subcategories, those weighing between 50 and 65 tons, grew 46% during this same five-year period, from 12,300 to 17,900. Overall, the number of heavy trucks, those over 13 tons, grew 37% from 1992 to 1997.[54]

Figure 13: Commercial Trucks Sold in Various Worldwide Markets

Year	United States	U.K.	Germany	Japan
1974	2 688 000	232 091	84 150	1 540 254
1984	4 093 000	222 895	106 419	2 320 938
1994	6 421 000	223 251	216 625	2 298 685
2003	9 357 000	363 687	264 745	1 363 833

Source: Ward's Motor Vehicle Facts & Figures 2004

Although these truck numbers are only a fraction of the number of cars on the roads, one eighteen-wheeler takes up the same amount of space as five cars when packed in a traffic jam, and two to four cars when moving at motorway speeds.[55] The end result is that as their numbers increase, they add disproportionately to the amount of congestion on the roads. It is not only traffic that increases; over 40,000 people are killed on European roads each year, and around 10% of those deaths involve heavy truck accidents. In one country, Sweden, heavy truck traffic has increased by 60% during the past 20 years, and now consists of 8% of road traffic, but trucks are involved in 22% of traffic-related deaths.

Why has the number of trucks on the roads increased so dramatically? Wal-Marting is a large part of the reason for this increase in the number of large and medium-sized trucks on the world's roads. What is "Wal-Marting", and what does it have to do with traffic congestion? I have made up the term to describe a logistics process that is one of the main reason's for Wal-Mart's huge success. It is so successful that it is being copied by most companies delivering goods all over the world, particularly Wal-Mart's direct competitors. The idea is simple: consolidate deliveries from suppliers, and optimize deliveries to stores. As with most simple ideas, the difference between successful application and failure is in the execution. It is like the *Toyota Production System*, which every automobile manufacturer tries to copy, but just can't seem to get the formula exactly right.[56]

Photo by the Author: The Wal-Mart logistics concept has been adopted and adapted across the entire trucking industry in North America and in Europe.

Sam Walton founded Wal-Mart in 1962 in Bentonville, Arkansas. Mr. Walton himself delivered all the goods to his first handful of stores in his station wagon.[57] His warehouse was his garage. At the time of his death on April 5, 1992, he was the second richest person in the world, after Bill Gates. In 1997, Wal-Mart became the largest employer in the US with 680,000 "associates" (everyone in the company, up to and including the CEO, is referred to as an associate by the company, rather than an employee). In 2002, Wal-Mart topped the Fortune 500 List as the largest company in the world in terms of sales, $218 billion. In 2004, sales rose to $287 billion and the number of employees worldwide had grown to 1.5 million at more than 3,600 sites in the U.S., and more than 1,500 sites in Mexico, Puerto Rico, Canada, Argentina, Brazil, China, South Korea, Germany[58] and Britain. Wal-Mart held the position as the world's biggest company until the end of 2005, when the high price of oil catapulted Exxon to number one, a shave ahead of Wal-Mart.[59] By 2006, the number of employees had risen to 1.8 million worldwide, with 1.3 million in the U.S.

Wal-Mart's success in attracting so many customers is based in large part on always delivering the lowest prices on all the goods it sells—not just on certain goods, or not just during special sales promotions, but *always*. They have perfected the art and science of penny-pinching, primarily by paying low wages to their staff, all of whom are non-unionized, and by squeezing their suppliers for lower prices and more donated labor. Because of their size, they can negotiate huge volume purchases, which drives down the prices they will accept from their suppliers to rock bottom. This, in turn, results in suppliers being forced to reduce their own costs of production. Fully 80% of the goods sold in Wal-Mart stores now originate in low-cost countries, mostly China.

The company has its own private fleet of 7,100 trucks and employs 8,000 of its own drivers. This puts Wal-Mart Logistics in a league just below truckload carriers Swift, Schneider and J.B. Hunt. (J.B. Hunt, the third largest fleet operator in the U.S., just happens to have its headquarters in Northwest Arkansas, right down the road from Wal-Mart's head office.) [60] The Wal-Mart fleet carries all goods to its Wal-Mart and Sam's Club stores, and, whenever possible, leverages its fleet by backhauling inventory from manufacturers' locations. This reduces the amount of contract shipping that is required, saving shipping costs. In fact, the suppliers pay Wal-Mart a fee for the service. The company reportedly saved close to $1 billion in freight costs in 2004 by backhauling inventory.

Goods are moved first from ports or producers' shipping points to one of Wal-Mart's 173 worldwide distribution centers. In the U.S, Wal-Mart has around forty general merchandise distribution centers that support stores within a 130-mile radius. There are another forty-or-so distribution centers for groceries that are within a 156-mile radius of the stores they service. There are twenty more centers for Sam's Clubs, seven for fashion, two for tires and one for e-retail fulfillment.

Rather than functioning primarily as storage facilities, as the traditional warehouse, there are cross-docking facilities at the distribution centers where goods are moved from supplier trucks to Wal-Mart trucks in the shortest possible time. Fifty percent of goods are cross-docked from delivery truck to a waiting distribution truck in less than twenty-four hours. The goal is to make just-in-time deliveries of exactly the right combination of goods to each individual store. And when the goods get to the stores, they are placed on shelves, not in storage.[61] The entire idea is to eliminate storage space, which is purely a cost for Wal-Mart, and to get goods into shopping carts and out the doors as fast as possible. High volume stores can have multiple deliveries and many shelf re-stockings per day. Warehouse space has moved from a fixed facility to moving vehicles: large tractor trailers; shelf-stacking carts; customers' shopping carts; and customers' vehicles.

This is where Wal-Marting adds to the entire traffic congestion picture. The greater the number of deliveries, the more trucks you will see rumbling down a highway and artery near you every day. These are not small vans, but large tractor trailers that are servicing many stores in a 150-or-so-mile radius from their home base.

What I have found most interesting about the Wal-Mart logistics solution is that it was the daughter of necessity. The key to the company's original success was that it located its facilities in poor, rural areas, starting in Arkansas.[62] Poor people shopped in their stores because they did not have much choice—if they were going to be able to purchase goods for themselves and their families in the same way that the more affluent were doing, as they were constantly reminded on TV, radio and in print advertising. Because Wal-Mart's facilities were not on high traffic routes, Wal-Mart initially had trouble getting its suppliers to deliver to its stores. This was the principal reason for the distribution centers, to provide more convenient locations for suppliers to drop off their goods. To get the goods to the stores from the distribution center, Wal-Mart was, in essence, forced to operate its own fleet of trucks. Once it owned the trucks, it could decide how to use them, and this is where the just-in-time delivery method and backhauling came into existence.

Let's look at how goods were delivered in the pre-Wal-Marting days. In the first half of the twentieth century, in cities of all sizes, households bought all of their groceries in stores within a five-to-ten minute walk from their home. There was a butcher store for meat. Chickens were bought in another store where they were picked still alive from their pens, killed and plucked while you waited. Boxed and canned goods came from a general store in the neighborhood. Bread came directly from the neighborhood bakery, or mothers baked their own bread. Deserts were homemade, often canned peaches, pears or cherries that had been put up in the autumn.

On Saturdays in the summer and autumn, farmers came from their farms outside of the cities in their horse-drawn wagons and sold their produce in each neighborhood. Horses were replaced by trucks in the mid-50s, much to the neighborhood children's dismay. Specialty products, like

Italian cheeses and cold cuts, Polish sausages, Jewish breads, were bought in local ethnic grocery stores. Milk was delivered daily to every house. One of the best treats for a youngster was to be invited by the "milkman" to ride in the milk truck on a hot summer day. Inside the truck it was refrigerator cool due to the large ice blocks that were placed in tubs under the milk racks to keep the products from spoiling.

These neighborhood stores had been in existence in most American cities since the time of major immigration, the end of the 19th and beginning of the 20th centuries. In the middle of the 20th century, very few people owned a car, and the idea of driving to shop for one's daily needs had not yet been conceived, at least not in most small corners of the world. The people who owned the shops lived in the neighborhood, often behind or above their shop. You bought from your neighbor, and if you had something to sell, they bought from you. Your neighbor's well-being was intrinsically tied to your own.

The shopkeepers had hand-carts or horse-drawn wagons, and eventually small motorized trucks, that they used to pick up their goods and produce from the *wholesale block*. There were few delivery vehicles at that time, except for the dairy trucks. *Wholesale blocks* were areas in cities close to the rail lines where produce and other edibles could be off-loaded and stored until they were picked up by the small retailers or distributors. They were messy, unsanitary places, which is one of the reasons why they have disappeared. But they served an important purpose. In the era of the small, local grocery store, they functioned as a staging point for food coming into a city. Oftentimes, the wholesalers had stands in front of their premises where they sold their particular products—usually the lower grades—to the public for very low prices. Every major city had a wholesale section. Some of the more famous are Le Halles in Paris and Covent Garden in London.

Slaughterhouses and meat packers were usually situated in locations close to the inevitable river running through the city. The same river also carried the raw sewage that was dumped into it from the sewers. Sewage treatment plants were not common until the middle of the 1970s.

There was a clear separation up until the 1960s between rural and urban areas, and life was very different in the two places. People who lived in the rural areas did so for a reason. They were farmers or they were providers of services to the farmers. In the midst of the farms there was usually a village with a small schoolhouse, a general store, a church (or, more commonly in U.S., multiple churches), and a farm supplies store. Shopping was definitely not part of the daily routine for those who lived on farms. The goal was to be as self-sufficient as possible. Major purchases were made during the infrequent trips to the nearest large city or from the mail order catalog suppliers, such as Sears Roebuck & Company or Montgomery Ward, both founded in Chicago in the heart of the mid-west farm and cattle country.

Small-scale, neighborhood shopping facilities, both in cities and in rural areas, did not require large numbers of deliveries per day. Consequently, they did not add significantly to the number of vehicles on the road. This began to change when both shopping and living patterns began to change, and fifty years later we are in the midst of the Wal-Marting revolution.

Wal-Marting has spread to other business arenas. Waste removal is one of these. There was a time when every community took responsibility for its own waste products, picking up its residents' trash and kitchen waste in dump trucks and hauling it to a landfill somewhere at the edge of their own city. In the mid-1960s, the price of landfill land exceeded the cost of shipping the city's waste to lower-cost sites, to other peoples' cities. These locations could be quite a long distance away, and making these long hauls was impractical for the standard city sanitary vehicles (a.k.a dump trucks). Wal-Marting was the solution.

The special-purpose sanitary vehicles would travel to transfer stations set up in convenient locations around the city. There, they would dump their loads. Large transport trucks would come to the transfer stations, and the waste would be reloaded onto these vehicles. The larger vehicles would make the long-haul trips to the distant sites. This solution, with several transfer stations instead of a single large landfill site, also had the advantage of keeping the sanitary vehicles in collection mode for longer periods of time because they could dump their loads at the transfer stations. Places like Northeastern Pennsylvania, with its highly developed road transportation connections to the New York and Philadelphia regions and its significantly lower land prices, became an ideal dumping site for large metropolitan areas.[63] It is a common site to see the large waste haulage trucks along the main routes between New York City and Philadelphia.

Another business area that has been Wal-Marted is package delivery. How many FedEx, UPS, DHL or TNT trucks do you see each day? UPS operates 88,000 vehicles of various shapes and sizes, TNT operates 19,000, and FedEx Ground has 18,000 motor vehicles. UPS has been in business since 1907, working both as a competitor and sometimes sub-contractor to the U.S. Postal Service. DHL is a wholly-owned subsidiary of Deutsche

Post World Net. It was originally an American company, started in 1969 by three men who gave their initials to the company name, Adrian Dalsey, Larry Hillblom and Robert Lynn. It was acquired by Deutsche Post World Net in 2002.

TNT is a subsidiary of Royal TPG Post, the national postal operator in the Netherlands. TNT is the global brand for express and logistics services. TNT Express has the largest integrated express air and road network in Europe.

FedEx Ground is a good example of how the delivery companies, and their vehicle fleets have grown. FedEx Ground began in 1985 as RPS (Roadway Package System), a division of Roadway Services, which became Caliber System Inc. in 1996. RPS revolutionized the small-package ground shipping market. It was the first in the ground business to use bar coding and automated sorting, providing customers with relevant information about their packages. In 1993 RPS exceeded $1 billion in annual revenue, just nine years after its creation, to record the fastest growth of any ground transportation company. By 1996, it offered 100% coverage of North America.

Following the acquisition of the Caliber companies by FDX Corp. in 1998, RPS was officially re-branded FedEx Ground in January 2000. Later that year, the company launched FedEx Home Delivery, a business-to-consumer service designed to help catalog and online retailers meet the needs of the residential market with standard features such as evening and Saturday deliveries. In September 2002, FedEx Home Delivery completed its expansion and is now available nationwide, serving to virtually every U.S. address.

In September 2004, Parcel Direct became a subsidiary of FedEx Ground after FedEx Corp. acquired this leading parcel consolidator. The service was later re-branded FedEx SmartPost and currently provides solutions for low-weight, less time-sensitive residential shipments. Today, FedEx Ground is the only small-package ground carrier operating a network of automated facilities, and ships more than 2.6 million packages every business day. It had sales of $4.7 billion in fiscal year 2005.

Trucks move the goods in most of the world today. The total shares of value and weight shipped by truck, rail, water and air in the U.S. have remained fairly stable for the past ten years, with truck shipments accounting for approximately 60% of both value and weight. Water shipments rank second in weight and value with around 15% and just under 10% respectively of the total. Air outranks rail in value for third place, while rail is in third for total weight shipped. Even in ton-miles, which is the highest weight shipped for the longest distance, truck shipments outpace both rail and water.

Figure 14: U.S. Commodity Flow Statistics by Mode – 1993-2002

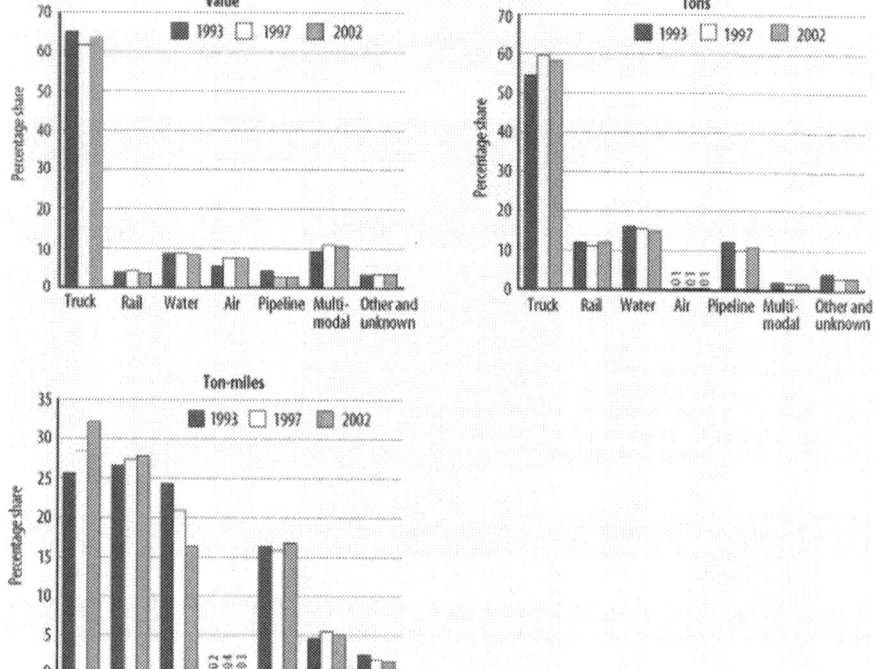

SOURCE: U.S. Department of Transportation, Bureau of Transportation Statistics, based on 1993, 1997, and preliminary 2002 Commodity Flow Survey data plus additional estimates from Bureau of Transportation Statistics.

NOTE: Multimodal includes the traditional intermodal combination of truck and rail plus truck and water; rail and water; parcel, postal, and courier service; and other intermodal combinations.

Shipping by truck is preferred because it is, overall, the cheapest, most convenient, most flexible and most effective means for moving goods in a Wal-Marted world. Although environmentalists would prefer that a greater percentage of total freight could, and should, be moved on rails and water[64], it simply is not as practical as it was fifty years ago. There is nothing on the horizon that is going to change that fact, barring a catastrophic event of an unthinkable proportion, like the total and immediate lack of available fuel.

There are even serious calls for longer and heavier trucks on the basis that it is more energy-efficient for one tractor to pull two or more trailers than the same tractor to pull just one trailer. Volvo Trucks and Scania, two Swedish truck manufacturers and among largest in world-wide sales of large trucks, have proposed that all European countries adopt the 24 meter standard length of truck allowed in Sweden and Finland, and drop the 18 meter maximum length restriction in force today in the other European countries. Their reasoning is that it will somehow reduce traffic, in spite of how counter-intuitive it sounds. They argue that it will reduce the total number of trucks by one-third, but it will also increase the total length of each truck by 25%. And road freight traffic will continue to grow, so that there will eventually be more and bigger trucks competing for space.

More and longer trucks will mean more traffic congestion. One way to reduce traffic congestion for commercial vehicle drivers is to get all the non-professional drivers off the roads. One way to reduce traffic congestion for private vehicle drivers is to remove all the commercial vehicles, especially the heavy trucks, from the roads. Unfortunately, at present there is only one road system, and it is shared by all drivers and their vehicles, commercial and private. The only way for you to beat the big truck traffic—until we implement the long-term fixes, like turning unused railroad rights-of-way into dedicated truck roads—is to get yourself off the road as often as you can.

Time to Get Unstuck

Now you know what causes congestion and, hopefully, while reading these pages, you have considered how it affects your life and the life of your family. Now it's time to take action. First, we need to take a look at how dependent you and your family are on driving to wherever you need to go. There are seven questions I would like you to answer. For each question, there are four possible answers. Associated with each answer is a number of points.

Circle all the answers that most correctly describe your actions. For example, if you have a teenage daughter with her own car that she drives to school, and a younger son who takes a school bus, select both answers. Add the scores for all four columns. The total for the last column, which involves no driving of any sort, is to be subtracted from the sums of the three first columns. So if you have two answers in the first column (10 points); two answers in the second column (6 points); one answer in the third column (1 point); and two answers in the fourth column (-10 points), your total score would be 7. The higher the number of points, the more dependent you are on driving.

If you have a score of under 10, you have been reading this book for its amusement factor because traffic congestion has little effect on your life. I hope you have enjoyed it this far. You won't need the advice in the next four chapters, so you can skip to the last chapter on how to take some of the pain of traffic congestion out of your daily drive.

If you have a score of over 25, you and your family are spending far too much time in your cars and, most probably, in traffic jams. The next four chapters will suggest ways to help those of you with high congestion quotients reduce your dependence on your vehicles for getting through the day. The recommendations focus on the four principal magnets for our daily journeys: school, work, recreation and shopping.

Travel Opportunity	5 Points	3 Points	1 Point	-5 Point
How my children get to and from school	We drive them or they drive themselves ☐	They take a school bus ☐	They take public transport ☐	They walk or cycle or No children ☐
How I get to and from work	Drive or car pool ☐	Park and ride public transport ☐	Take public transport ☐	Walk or cycle or work at home ☐
How my spouse gets to and from work	Drive or car pool ☐	Park and ride public transport ☐	Take public transport ☐	Walk or cycle or work at home ☐
How I and/or my spouse get to and from shopping	Drive ☐	Park and ride public transport ☐	Take public transport ☐	Walk or cycle ☐
How I or my spouse get to and from weekly recreational activities	Drive ☐	Park and ride public transport ☐	Take public transport ☐	Walk or cycle ☐
How my children get to and from daily recreational activities	We drive them or they drive themselves ☐	Another parent or friend drives them ☐	They take public transport ☐	They walk or cycle ☐
How I and my family get to weekend recreational activities	Drive ☐	Park and ride public transport ☐	Take public transport ☐	Walk or cycle or stay at home ☐
Vertical Score	☐ +	☐ +	☐ +	- ☐

Total Congestion Quotient	

Let the Kids Walk to School

Why does it seem like our dream of freely flowing traffic comes true whenever there are school holidays?

DID YOU WALK TO SCHOOL AS A CHILD? Perhaps you are old enough now to have attended school at a time when your parents would not have been considered irresponsible for allowing you, as a grade-schooler, to walk to school completely on your own. You could move at your own pace, as long as you arrived before the bell stopped ringing and the entrance door was locked. In the autumn, you could fill your pockets with horse chestnuts that had fallen from the trees and were freed from their prickly husks. You could form the perfect snowball in the ten minutes it took your small legs to travel the distance from your front gate to the school yard, and test your pitching arm by trying to hit a telephone pole at ten paces, making sure that you did this before a teacher or the school principal could catch a glimpse of you. Maybe you enjoyed meeting up with your friends along the way and talking about who would win the World Series or the FA Cup, which new car model you hoped your parents would buy, or what was the best rock'n'roll song on the charts.

For a combination of reasons, children do not walk to school anymore, either alone or in groups, in the same numbers as they did forty of fifty years ago. One of the main reasons, which I discussed in an earlier chapter, is that schools are just too far from children's homes to make walking practical. Schools increased in size and collected students from larger and larger catchment areas between 1940 and 1990. While the population of the U.S. grew by 70% between this period, the total number of elementary and secondary schools fell by fully 69%![65]   In a 2001 study by the U.S. National Household Travel Survey, researchers found that fewer than 15% of students in the U.S. between the ages of 5 and 15 walked to and from school, and only 1% bicycled. In 1969, 48% of students in the same age group walked or bicycled to school.[66] Another U.S. study showed that in 1999, only 31% of students in the 5-15 years old age group who lived within one mile of school walked or biked.[67] In 1969, close to 90% did so.[68]

If children are no longer walking or cycling to school, how are they getting there? Depending on their age, the vast majority, fully 70% in the U.S., are either being chauffeured or they are driving their own cars. Using the word "chauffeured" is deliberate. There was a time when only the rich could afford to have chauffeurs, people who were hired only to drive the

Rolles Royce, Cadillac, Mercedes or Jaguar to wherever the wealthy family members needed to be transported, including school for the children. Today, wealthy or not, children are driven to where they want or need to go, and school is one of the places on their route.

The number of children taking the classic "Yellow School Buses" in the U.S. was under 15% in 2004 and dropping. These buses were rare in urban areas until the 1970s. Before that they were used in cities for picking up children with special needs and taking them to their schools. They were used in rural areas where naturally long distances between farms and villages made them a necessity if children were going to go to school at all, since farming families had little time to chauffeur their children. The yellow buses appeared in many U.S. cities as a result of court rulings to integrate the schools. While de jure segregation in the south had been ruled unconstitutional in 1954, de facto segregation continued to exist inside and outside the southern states.[69] Whether it was the result of income differentials and location choices, or discrimination in housing sales, the consequences were that the racial makeup of the nation's schools tended toward homogeneity: white children went to schools where most of the other children were white, and black children attended schools where most of the children were black.

The courts ruled that this had to change. Children would be bused from one neighborhood to another so that each school reflected the racial mix of that city and region. Parents protested. Teachers protested. Even students protested to being taken from their local schools and forced to attend classes in distant neighborhoods. No one seemed to give a thought to the traffic consequences of busing children around a city region. If they did, the issue was viewed as secondary to the main goal of integrating the nation's schools.

Driving the kids to school is not a U.S.-only phenomenon. The following is an estimate of the problems caused by children being driven to school by their parents in the United Kingdom:[70]

- Parents in the U.K. spend a total of 570 million hours per year on the school run, which is about one hour per day per accompanied child;

- Other commuters lose a total of 130 million hours per year due to increased congestion that would not be there if parents were not driving their children to school;

- An additional 500 million liters of fuel is used, resulting in 2 million tonnes of carbon dioxide emissions; and,

- Each year accidents on the school run kill 40 children, seriously injure 900, and slightly injure a further 6000.

At the time that low-density suburban communities were being built, someone could have stood up and said: "The children must be able to walk to school." That would have meant a completely different approach to both school buildings and teaching than what has evolved. Instead of large facilities with many teachers and administrators, and with special-function spaces, we would have built small, intimate structures that fit into the residential community, with a small number of teachers and staff. Instead of facilities surrounded by parking lots, and entrances designed for bus and car drop-offs, there would be a minimum number of parking spaces, probably on the street, and no allowance for student parking. Instead of expensive special-function spaces, each facility would be multi-functional.

Not all children who are driven to school prefer this alternative. In a U.K. study, 38% of the children driven to school said they would prefer to walk or cycle.[71] Why? They did not like being stuck in traffic; they did not like cars causing pollution; they did not like the fact that they were not getting proper exercise; they did not like not meeting their classmates; and, driving was simply boring.

It is a fact that no one did stand up back then (or if they did, no one was listening), and they did not ask the children what they would prefer, so the small community grade school, middle school and high school were not built in the new suburban developments. Neighborhood schools have even begun to gradually disappear in our cities, as the figure cited above on the reduction in the total number of schools testifies.

There is another factor that is creating a separation from where a child lives and where that child attends school. It is call *school choice*. One means for providing this choice is to offer parents *school vouchers*.[72] With a voucher, a child can attend any school within an area designated by the school authorities. The idea is that every child can choose his or her school based on a number of criteria, such as closeness to home, existence of desired facilities, availability of desired courses, qualifications of the teachers. Another motivation is the desire to improve education for the economically disadvantaged. A child is a poor neighborhood with low quality facilities can attend a high quality school in a wealthy neighborhood. The low quality schools, or the ones with sub par facilities, inadequate programs or disinterested faculties, will have to improve or "go out of business".

In order for a voucher system to work in practice, the children must be able to travel to their schools of choice. If they cannot walk or cycle, or if there is not a transportation system that can take each child to every school, then they must be driven. And this adds to the traffic congestion problem.

What are some of the other reasons, besides distance, given by parents and students for why children no longer walk to school:

- There are no sidewalks, or there are too many dangerous street crossings for children.

- There are fears of pedophiles.

- There are fears of other criminal elements, like drug pushers.

How can we, in good conscience as parents, send our children out into this jungle—alone? We cannot. A study in the U.K. of school travel patterns reported that one of the principal reasons parents drive their children to school is that they feel it is a sign that they care about their child's safety and well-being. They feel guilty if they do not drive them, and fear that other parents will think of them as uncaring parents if they allow their children to travel unaccompanied to school.

So, on the one hand, lower housing densities and fewer, larger schools have created the need to transport children to school, while on the other hand, fears for the safety of our children has fostered a mentality that the only way to have peace of mind is to drive our children door-to-door to school. That's the only way we can be sure they are safe. It's a vicious circle that we have created. The end result is increased traffic congestion during the morning when adults are also trying to get to work. A by-product is an increasing incidence of child obesity, which is increasing all over the world in parallel with adult obesity.[73] We can help to solve both problems by stopping the morning school run.

The first step is the most important. You can decide where to live based on the shortest and safest walk to the nearest grade school, and the nearest trip by bicycle to a middle school and high school. You can set this at the top of your priority list when deciding where you will settle or re-settle your family. If children are going to walk to school, it must be possible for them to walk on safe paths, not on the road. Choose a place where there are sidewalks. Places where there is multi-acre zoning and horses behind white rail fences tend not to have sidewalks. If you aspire to live in a community such as this, or already do, then you probably are thinking you will earn enough money to send your kids off to boarding school and won't need to worry about driving them yourself each morning. Or maybe the nanny will do the chauffeuring. In any case, you are probably not that concerned about this part of the traffic congestion problem. For the rest of us, we should think sidewalks. Outfit your children with good pairs of walking shoes, proper rain gear, and warm clothes for the winter if you live in northern climates, and do the same for yourself, and get out there and walk.

One positive step you can take is to vote for accessible schools in your community. Schools should be small and dispersed throughout the community. The *Bill and Melinda Gates Foundation* has donated $1 billion over five years to create 1,500 new small high schools. The stated aims of this donation is to make schools accessible to children by walking or cycling, and to re-connect the school to the child's neighborhood. The U.S. Congress appropriated $125 million to the Smaller Learning Communities program for fiscal year 2001, and a further $142 million for fiscal year 2002 in order to assist communities in downsizing their schools.

There are no quick fixes to the problems with predators. We can, as parents, take turns walking to and from school with our children. Another option is being tested in the Gothenburg, Sweden region. It is a walking school bus. The idea is based on the premise that there is safety in numbers. A group of children walk to school on a selected route each day. Children gather at collection points along the route and wait for the "bus" to arrive. As they join, the "bus" grows larger. The "bus" will wait a few minutes at each collection point for stragglers, but the goal is to keep to the schedule. Results have been very positive, with the children stating that they prefer it to being driven to school. They like the exercise, meeting their friends, but most of all, contributing to a greener environment.

There is an international movement called *International Walk to School*. A week in October is Walk to School Week. The idea actually started in the U.K. and was carried over to the U.S. in 1997. A group called *Partnership for a Walkable America* sponsored the first *National Walk Our Children to School Day*. It was intended to support walkable communities, with safe streets and crossings. By 2002 it had expanded to all 50 states and to other parts of the world. In 2003, *Safe Routes to School* (SR2S) legislation was introduced in the U.S. Congress SR2S programs may include policy development, planning and implementation of strategies such as improvements to streets and sidewalks, education and encouragement of children and parents, and increased enforcement of traffic laws. As described on the organizations web site, programs can include:

- Walkability and bikeability audits of the safety of streets around schools;

- Local programs to improve sidewalk conditions near schools;

- Use of traffic calming devices to slow traffic and give pedestrians priority;

- Programs that educate children on walking and biking safely, and challenge them to walk or bike often;

- "Walking school buses" in which one or two parents or volunteers escort a group of children on the walk to school;

- Increased traffic enforcement around schools;

- School construction that includes renovation and improvement of existing schools, and locating new schools to reduce walking hazards and avoid major traffic threats; and,

- Cooperation between school officials, law enforcement officials, and transportation planners.

Fifty years ago there were the eyes and ears of adults who would look after their neighbors' children as they walked to and from school each day. The eyes and ears belonged to mothers who were not working, retired grandparents, shop keepers, delivery drivers, or even the older children. Today, both parents are working—because they have to or because they want to—grandparents are in nursing homes or retirements communities, separated from daily activities, shop keepers and the delivery van drivers who serviced them have disappeared, Wal-Marted. Maybe the only solution is for children to look out for themselves. Starting and ending their day with a drive in the family car is not really very good for anyone, not for them, not for their parents, and not for society.

Separate Transportation from Recreation

THE SOCCER MOM PHENOMENON didn't always exist. It started in the late '70s and early '80s in the U.S. It began with the appearance of the cute window sticker proudly stating: ***Baby On Board***. I was never sure what to think about those signs. Should I be happy for the new mother, or should I be offended by the insinuation that I was not normally a careful driver who was considerate of those with whom I was sharing the road, irrespective of whether they were carrying a baby or a mother-in-law or a poodle or just a load of groceries? In the end, I decided to choose the benevolent alternative, to be happy for the new mother. But it was a bit sad, I thought, that she needed a sign to advertise her new baby. She wouldn't need it if she were walking with the baby in a pram, rather than having the infant barely visible inside her car, packed into another new addition to the automobile, the child safety seat[74].

Driving the "baby" continued through nursery school, grade school, junior high and high school years, and not just to school, but to all the activities that can loosely be called *recreation*. **Re·creation** is an activity that refreshes and renews one's health and spirits by enjoyment and refreshment.[75] The origin of the word is the Latin *recreatio*, meaning "to restore to health", which is created from the Latin *re* meaning "anew" and *creare* meaning "create". So recreation literally means "to create oneself again".

At some point, recreation ceased being a spontaneous activity for children and became a set of contrived activities organized by adults. Something is **spon·tan·e·ous** when it proceeds from a natural personal impulse, without effort or premeditation. It was as if adults decided that playing was a waste of a child's time, that if it wasn't recorded, it would not count on the child's eventual application to university; that if it did not take place in a purpose-built facility, the sports scouts who were constantly prowling for future talent would never see their children. Empty lots with their grass scythed by helpful adults, or even by the enthusiastic children who used them, disappeared, replaced by large, well-clipped playing fields. Frozen ponds were abandoned in favor of enclosed ice rinks. Schoolyard basketball courts were left to fall into disrepair, and neighborhood baseball fields lay unused.

Safety has been the main justification for this transition from spontaneity to premeditation. Children are safer when they play under the supervision of adults, claim adults. They don't need to worry about being drowned falling through holes in the ice, twisting their knees and ankles on uneven surfaces, or impaling themselves on picket fences reaching for a long fly ball. They learn the official rules, so there are fewer disputes leading to fights, say the parents (although sideline parental bickering or fistfights are now common). They are instructed in the proper techniques, so they don't develop any performance hindering habits. They play on regulation-sized courts and fields and rinks, so they are prepared to compete at all levels up to being professionals some day if they are good enough. These are the arguments given in favor of non-spontaneous play.

It is difficult to deny all of the advantages to organized play, but "official" facilities are usually centralized because they are expensive to build and maintain, and they need to be shared by as many people as possible. The result of this movement away from spontaneous play and small neighborhood facilities, which was well underway by the beginning of the 1980s when the Baby Boomers' children were getting old enough to recreate, was more driving to get to these central facilities, and more traffic congestion along the way.

Quite a bit of that driving was done in a new type of vehicle, the *minivan*. This was a car type invented by Chrysler Corporation (now DaimlerChrysler). Chrysler engineers were given the task of designing a station wagon-type vehicle that was not derived from another vehicle, such as a passenger car sedan or a commercial van. At the time work began on the *minivan*, Chrysler dominated the full-sized van market, with a market share of 45%. They were successful with the van because they offered car-like conveniences, such as power windows and locks, quality audio systems, and rear window defrosters. They felt they could create a smaller format van for families, one that would not compete in the same segment as their full-sized van, but would instead steal sales from station wagons, a segment in which Chrysler was not at all strong.

Chrysler had started working on their minivan concepts in the early 1970s, and full development started in 1980. The Dodge Caravan and Plymouth Voyager were first introduced in 1983 as model year 1984 vehicles, just about the time the Baby Boomer moms and dads were starting the trend of chauffeuring their children everywhere, and taking turns being the driver for their friends' children. Applauded by automotive magazines, the minivans became a major success from their debut, with a total of 209,895 sold in North America in 1984.

One of the key features of this new vehicle type was the sliding side door. This was based on consumer research performed by Chrysler. It showed that customers felt it was safer when getting children in and out of the vehicle. A sliding door would not blow closed, it would be less likely to trap fingers, and it provided a large access opening.

Both General Motors and Ford introduced their own minivan versions in 1985, the GM Astro and Safari, and the Ford Aerostar. Models from Toyota, Mazda and Honda followed soon afterward. It was a decade later that the minivan began to make an appearance on roads in Europe, for the same reasons as in the U.S.

Chrysler Corporation's Dodge, Plymouth and Chrysler minivans became the transportation workhorse for baby boomer families in the 1980s.

The minivan became the family's second car, and it was used to ferry the kids to all of their pre-school and post-school day activities, such as the four a.m. ice hockey or figure skating practice, to the after school ballet, piano or karate lessons. We began driving our children to their play in the same way and for similar reasons as we began driving them to school: Fears for their safety; too long distances to walk; more convenient for the parents. And we have continued. Between 1969 and 1990, the U.S. average daily vehicle trips taken for recreational purposes increased 67%. The **Baby On Board** sign eventually gave way to a **Mom's Tax** sign.

In time, the vehicle changed from a minivan to an SUV (Sport Utility Vehicle). Chrysler Corporation led the way in this category as well with its re-designed and downsized Jeep. There then followed a lineup of SUVs to fit every pocketbook and size requirement, from the small format Toyota *RAV4*, to the jumbo-sized Chevrolet *Suburban*, from the diminutive Land Rover *Freelander* to the oversized, in-your-face Ford *Expedition*. Fuel was relatively inexpensive in North America where the SUV explosion took place. In 2001, sales of so-called light trucks, that include SUVs and pick-up trucks, for the first time exceeded sales of automobiles in North America, 8.7 million to 8.4 million.[76]

The baby boomer's children were growing up, and their recreational possibilities were finally beginning to coincide with the dormant interests of their parents, who had put their skiing and hiking and white water canoeing on hold while their infants turned into small adults. Minivans were fine for the school run and for scooting around the suburb, but for the weekend escape and vacation excursions, for pulling the boat, the horse, or the camper, and carrying all the gear along with the family, a real utility vehicle was needed. Automakers were happy to oblige.

Adults have been adding to the traffic congestion problem, getting themselves to their tennis, racquetball and squash matches, to the gym, to aerobics, to line dance get-togethers, to tango and samba lessons. It seems that people of all ages cannot separate recreation from transportation. On the one hand we are being urged in media campaigns to exercise in order to stay healthy, but the only opportunities to engage in exercise seem to require that we get into a car and drive or be driven. One of the reasons that we do not have the recreational opportunities we would like to have close to our homes is that we spend so much time escaping to places where recreation is packaged for us. Time shares in the Rocky Mountains and on North Carolina beaches, villa rentals in Provence and Tuscany, beckon us to leave our own communities to add to the economies (and congestion) of someone else's. Why not stay at home and make your own community a place where you would really like to live, work and play?

Walking and cycling are part of the answer to this problem. Instead of driving for half an hour to the gym, exercising for an hour, and driving half an hour back home, take an hour walk by yourself, with a family member, a friend and/or your dog. Add a pair of ski poles to give your upper body a workout while walking, and you have a perfect exercise, one that is less damaging to the knees and hips than jogging on pavement. You get some social time if you are with someone, or some good thinking time if you are on your own. If you have a dog, he or she will introduce you to your neighbors, broadening your social network. You will even save time and maybe get a better workout than at the gym. And on top of all this, you will also add a pair of eyes to your neighborhood, which is always good for reminding would-be wrong-doers that they are watched.

Bicycling has long been recommended by doctors and physical therapists as an alternative to the more body part punishing sports, like jogging, squash or even swimming and aerobics, especially for people suffering from chronic back pain.[77] According to the U.S. National Institutes of Health, nearly four out of five people will experience back pain at some time in their life resulting from a wide number of causes, particularly excessive physical activity or effort, improperly warming up before a physical activity, and poor flexibility.

The only valid argument that has been used against the neighborhood exercise routine is fear of personal attack. We're back to the safety issue again, where we were with kids walking to school. We don't walk or run or cycle because it's not safe, and it's not safe because there are not enough people out there keeping a watchful eye out for their neighbors. It's a classic Catch 22, chicken-and-egg dilemma. Well, we know that the chicken came first; it's just a matter of figuring out which one is the chicken and which is the egg.[78]

New York City's Central Park is a prime example of how a recreational opportunity can be wasted by fear, and how it can be brought back to life by good will and cooperative hard work. By the end of the 1980s, New York City had become a city where outdoor recreation consisted of drug running and running from muggers. The city had spiraled down from being a livable, healthy and cohesive city in the first half of the 20th century, to becoming a cesspool of moral and physical deterioration in some areas and extravagant wealth on exhibit in others. In the middle of these two extremes, the majority of the city's residents tried to cope and get on as best they could with their lives.

Few places symbolized this sad state of affairs more than Central Park, where the edges along the high-rent districts were kept neat and tidy, but inside it was a no-go zone. The mayor of New York City at the time, Rudi Giuliani, reasoned that the "chicken" in the fight to win back the city from the dark forces was getting rid of crime, and he and his administration set out to do just that after they came into City Hall in January, 1994. They established a zero tolerance policy toward anti-social behavior, everything from public drunkenness to graffiti. They focused on Central Park. Step by step they won back ground until one day the people of New York City declared that The Park was theirs again.

There is no excuse for citizens in any democratic society to live in fear. The primary role of government is to secure the safety of its citizenry, to protect its people against destructive forces. If you cannot walk, run or cycle around your neighborhood because you are afraid of being assaulted, then there is something very wrong with the government that is running your town. Get your politicians to start doing their jobs. There is not a fast financial return to be made from increasing personal safety, like there is for charging drivers for entering the city, but the long-term positive effects to the community as a whole from a safe and secure environment will be much greater. If your mayor or city governing council is not promoting safer streets and parks, maybe it's time you to raise your voice and suggest that they do.

Shop Locally

IT IS NOT EASY TO MAKE THE TOUGH TRADEOFFS between driving to a major shopping center or mall or building supply depot, and frequenting a local supermarket or variety or hardware store. We have grown accustomed to maximizing our choices and minimizing our purchasing costs. The car has given us the freedom to go where the goods we want are sold, and worldwide supply chain management delivers those goods to us at always the <u>lowest price</u>. There still are places on the planet where primitive civilizations continue with a life of hunting and gathering, making the tools they need to survive. What they cannot make, they don't need— or at least they don't know that they could use them because no one is bombarding them with new product advertisements. There also are places where the earliest forms of walled cities exist, where daily life consists of making things that are traded for life's necessities. A visit to the *medina*[79] in Fez, Morocco, for example, is an enjoyable journey back one thousand years in time. In Fez, as in the ancient cities, the streets are wide enough for a donkey and a small cart pulled behind. The shops where goods are made and sold sit tightly packed along these narrow passageways with the shop owner's family quarters usually located on the floors above. Farmers and herders come to the city and set up their temporary *shops* in the large open area outside the city's gates. The city dwellers come out to buy produce and other wares, and the country dwellers enter the city to purchase what they need and can afford. This type of life exists today in many corners of the world: in Asia, in the Far and Middle East, in Africa, in South America. These places are few and diminishing. Modern communications and energy generation methods have brought the global, consumer society and shopping to even the most remote corners of the earth.[80]

In the rest of the world, where cars and trucks and busses and trolleys compete for space with pedestrians, and where traffic congestion exists, small shops operated by specialists in one product or service, such as the meat, fish or produce monger, clothes tailor or shoemaker, eventually grew in size to department stores and super markets. The stores moved from the centers of cities to large, open fields at the cities' edges. Eventually, the "life's necessities" shops were collected into malls, and small, specialty shops filled in the spaces between the larger variety stores. Today, the malls, and all of their individual stores with all of their contents, have been consolidated under one roof into the sell-everything-at-always-the-lowest-price, super-sized, "Big Box" stores, like Wal-Mart and Costco and Carrefour.

At the same time as this consolidation and concentration was occurring, the smaller, more locally accessible stores were disappearing. It has gotten to the point in most parts of the U.S. that even if one wanted to shop locally, there would be nowhere to go, and those stores that are close by do not have most of what we want to purchase. In Europe, where there remains a strong tradition of living, working and shopping in cities, the same forces are eroding the base of shops. In Stockholm, Sweden, for example, the number of everyday commodity stores per 10,000 inhabitants shrunk from 8.3 in 1975 to 6.2 in 1990 and further to 4.8 by 1999. In the period from 1975 to 1999, the number of large stores, with sales over SEK 125 million ($18 million) increased almost 300% from 15 to 44, while the number of small shops with less than SEK 60 million ($8.5 million) contracted by 40%, from 1,202 to 744. This is particularly troublesome for the elderly who cannot drive, and for those others who, for whatever reason, do not own or drive a car.

When I began researching the topic of traffic and shopping, I was surprised by how much is being done by small, independent businesses in North America and in Europe to fight back against the large national and international chain stores. One group is the American Independent Business Alliance (Amiba) based in Bozeman, Montana in the U.S. "Unlike global corporations," said Jeff Milchen, one of the founders of Amiba, "local businesses are owned by people who live in the community and are committed to its well being. These businesses are vital to our quality of life and sense of place, but they face powerful threats, and it requires conscious action to ensure their endurance. Thankfully, many communities are responding to the challenge, realizing that the best community qualities don't come in Big Boxes."[81]

In addition to threatening local businesses, "Big Box" stores like Wal-Mart, Costco, Home Depot or Dollar General create monstrous, unsightly premises surrounded by vast, paved parking lots. These chain stores are prone to abandoning their facilities on a regular basis to build even more monstrous, unsightly premises with more parking in the vicinity of the old store[82]. More to the point for our traffic problems, they generate two kinds of traffic congestion: they are the destination for a continuous stream of large trucks making frequent deliveries; and they attract large numbers of cars entering and exiting the facility's site to fill their parking lots.

If you shop at one of these stores, you are both a victim and a part of the problem. If your aim is to minimize the inconvenience, irritation and time delays that traffic congestion has on your life, then your objective should be to avoid shopping in these facilities. This is, perhaps, easier said than done.

Like making other congestion-avoiding decisions, we may have judged that the extra costs of travel and the irritations of delays are more than compensated for by the benefits we gain as a result of the journey. In the case of shopping, the benefits are the calculated savings achieved and the added certainty that we will find the products and services we both need and desire. We could shop in a store that is a short distance from our home in the center of a town or city, one that is accessible by walking or by public transport, or we could travel to a "Big Box" store. We choose the "Big Box" store because we make a comparison of the costs of items in the local store with the same items in the "Big Box" store, and the latter wins hands down for the reasons I stated in the earlier chapter, *Wal-Marting the World*. These large stores have commoditized the products we purchase, meaning that through their purchasing power, they have reduced the product's cost to a considerably lower level than is possible in stores that do not have such scale advantages.

Local, family-owned establishments of all kinds, from lumber yards to hardware stores, from bakeries to clothing stores, are closing their doors because their former patrons make this simplified calculation and choose to shop at the national and international chains, instead of at their neighbors' stores. Why bother to visit the local bakery if all one's bakery needs can be filled at the same time one is shopping at the Big Box store, and at a substantially lower cost. Once products are commoditized, the only differentiating factors between stores selling the same basic products are <u>quality of service</u> and <u>variety</u> of other types of products. "Big Box" stores can deliver the absolute lowest prices and broadest selection of products because of their sheer size and purchasing power. Chains such as Home Depot have institutionalized service by making the provision of professionally trained personnel in their stores a key part of their offering, or Wal-Mart can give the illusion of service by having cheerful employees constantly greeting the customers with friendly "Hellos".

We consumers, therefore, put up with the congestion because we believe we save large amounts of money, have the widest possible product choice, and are served by cheerful and knowledgeable staff. Is this really the case?

Let's look first at the issue of lowest price. It is worth asking the question: Is buying at the lowest price always in my best interest? All things being equal, why pay more for the same product? This is one chain's motto: *Why pay more?* they ask in ubiquitous advertisements. The implication is that only fools pay more when they can get the same product for less. If the amount of money a household has to spend is fixed, it makes good sense to maximize the goods and services that can be bought with that money, doesn't it? But there are some serious and convincing arguments that have been made against this economic self interest. One of these arguments, using Wal-Mart as the case study, is not so obvious to the individual shopper. Shopping at these low-price stores may cost you or someone close to you your job.

The way that Wal-Mart achieves its "everyday low prices" is by establishing a benchmark price for a particular item they will sell in their stores, such as a box of dish washing detergent, and then expecting all suppliers of that item to meet this price. If a company wants the big volume business that Wal-Mart can deliver, it accepts this price. If the company believes that it can continue in operation without a Wal-Mart's contract, then it simply refuses to sign up to the price. For those who do become suppliers to Wal-Mart, they are forced to find ways to reduce their own costs in order to meet the increasingly low price points set by their demanding customer. This has meant finding lower cost raw materials and manufacturing opportunities. When a company has squeezed all of the cost savings it can out of its local labor force and sub-suppliers, and still cannot get down to the level that would result in a profit, the only alternative is to move material sourcing and production offshore to a low-cost country, like China or India.

It is no accident that 80% of the products now sold in Wal-Mart stores originate in these low-cost countries—even though the company whose brand is on the product is American or European—because suppliers have found that this is the only way to meet Wal-Mart's price demands.[83] And it is not only Wal-Mart. It is every company selling in the global market. We read the headlines about one business after another moving their production "off-shore", and shake our heads about foreign competition taking our jobs. But we don't always make the connection between how our own actions, like buying the lowest price option, results in the effect of someone—including ourselves—losing their job.

A look at Italy indicates what this type of low-price sourcing can mean for a country's economy and its workforce. Between 1995 and 2005, Italy lost 600,000 jobs in textiles and clothing, which is about one-third of the sector's total in the old 15-country European Union, and one-quarter in the expanded 25-country area.[84] In 1990, Italy's global clothing manufacturer, Benetton, bought almost 90% of its stocks from home market companies. Today, that number is down to less than 30% and it is expected to drop to below 10% during the next few years.[85]

So far the winners in the era of globalization and outsourcing are companies and the top wage earners. Company profits as a share of GDP increased in the U.S. from 7% in 2001 to 13% in 2006, and the country's top 1% of wage earners received 16% of all income, compared to only 8% in 1980. Real wages for the middle income worker has fallen by 6% since 2000. The Economist, a perennial supporter of free trade and globalization, has had to admit that globalization may not be all that they cracked it up to be, with the pie growing bigger for everyone, and wages and jobs growing in both the industrialized and emerging markets.[86] Wages will stagnate or decline, especially for those who must compete with offshore companies, and while jobs will be available, they will be lower paying. And wage growth in sectors that cannot be outsourced, like handing a hamburger over the counter or greeting customers with a big "Hello. Have a nice day." will be controlled by a seemingly never ending supply of new arrivals to the country's shores.

What has happened to the workers in North America or Europe who formerly produced the products prior to their outsourcing to low-cost countries? They are either unemployed, in a re-training program, working for the next company to be Wal-Marted, or working for Wal-Mart or one of its competitors. They are most likely working for a lower wage, probably in a non-unionized business, and trying to save as much money as possible on the purchases of their household needs. If they did not feel compelled to shop at Wal-Mart stores before they were forced out of their former jobs, they certainly do now.

Isn't this just the way of the industrialized world? North America and Europe had their time in the sun as the places where goods were produced. We took the raw material and the immigrant labor (free-willed or forced) and built mills and factories on top of these resources. We prospered as countries, and the granddaughters and grandsons of the immigrant coal miners, shoemakers, dock workers, farm laborers and factory workers are today part of the economic, social and political mainstreams in their countries. In the U.S., jobs moved from North to South, from East to West, and eventually to Mexico and other off-shore locations. South

America had an opportunity to shine but lost its spark. Now it's time for Asia, first with the "tigers", South Korea, Hong Kong, Taiwan, Thailand, Singapore, Malaysia and Indonesia, and now with China, India, Vietnam and Cambodia. One day it will be Africa or the Middle East, maybe Pakistan and Bangladesh. Perhaps, in time, the low-paying jobs that are the engine of industry will return to North America and Europe. I am not sure that this is the way things have to be, but I do believe that people should be aware of the choices they are making, either knowingly or unconsciously.

What about selection? Many of the "Big Box" stores have introduced their own brands. They sell these products at prices lower than the name brands, and generate much higher profits from their own brands than from those sourced from the brand manufacturers. Why not just load up the stores with their own brands? The reason is variety. Shoppers like variety; they like being able to compare prices, and at least knowing that they have made a choice between different products. Some choose only on the basis of price, and they buy the store brand, while others choose based on other criteria, such as known quality, taste preference, image, and they purchase the known brand.

There is often nothing distinguishing the store brand from the known brand. The products are often manufactured in the same outsourced plants in low-cost countries. The difference is the name on the package. However, in many cases, the store brand does not have the same look, feel or taste as the known brand because the store's supplier cannot make an exact copy of the known brand's product. The recipe may be a well-kept secret, like Coca Cola, or the design may be patented.

Against this onslaught, the local hardware, appliance, dry goods, clothing, music or grocery store has very little effective ammunition with which to counter attack—except a central, community-based location, and possibly a personal face, real product knowledge, and a history of participation in the community. Prices will be higher in these shops; that is a given. However, if we do not choose to shop in them, they will disappear altogether. It is already happening. Visit any small or medium-sized town to see examples of it. If there is a clothing store, it will be hanging on by its threads, is probably being run by an elderly man or woman who have the store in their blood, and they have little chance of selling it or passing it on when they finally retire or expire. The selection of clothes is likely to be outdated, maybe even ancient. The grocery and hardware store shelves are probably stacked one box deep with minimal selection.

What can we do, if anything, to avoid being forced to get into our cars to satisfy our most minimal shopping needs? Unless we are willing to make some drastic changes in our lifestyles, like moving to Fez and becoming a copper pot maker, the best we can do is to try to support any and all of the local shops that remain in our neighborhoods, and, at the same time, work for long-term changes in the design and location of shopping facilities. Visit the local stores as often as you can. Buy something. Anything! Bring a friend and have them buy something as well. You may actually find a flannel shirt or a pair of those old style of jeans—you know, the ones that came up to your waste and had back pockets also up there where they should be—that were your favorite but you can't buy any longer in one of the chain stores. If you do find a place that does carry more up-to-date styles, they are likely to be a bit more expensive than at one of the stores in the mall. Pay more, buy less, keep your neighbor working, and possibly save your own job. In the bargain, you can avoid the traffic at shopping centers and do your part to cut down the amount of truck traffic traveling into your region.

Most corner grocery stores are gone, but there are still plenty of smaller supermarkets in operation in most medium- to large-sized communities. They often have a specialty along with the standard fare of produce, meats, boxed and canned goods and sundries. Shop in these stores rather than in the giants. Set the budget for your total purchases at the level you think you would pay for a week's worth of groceries bought at the Big Box, and throw out of the cart the extras that break the budget. Remember, it's a tradeoff, and on the plus side are less driving and less traffic.

Now to the changes that you can support. Encourage your community leaders to eliminate single-use zoning codes in favor of multi-use developments. Zoning was a noble idea when it was first introduced in New York in 1916 as the city's reaction to a skyscraper overshadowing neighboring residences. They wanted to prevent similar occurrences from ruining the residential areas of the city. Eventually, zoning regulations were used in the U.S. and in Europe to separate theoretically incompatible land uses into isolated packages of industrial, retail, commercial, recreational, institutional and residential. The theory was to promote healthy environments by keeping the places where people lived as far away as possible from the toxic air generated by manufacturing, the filth of food processing, and the clamor of goods exchanging. In its ruling on the challenge to zoning laws, the Supreme Court ruled that "...a community may enact reasonable laws to keep the pig out of the parlor, even if pigs may not be prohibited from the entire community".

It would have been better to attack the source of the problems—clean up or get rid of the pig—and to have kept everything in proximity to everything else. Through zoning, we have ensured that we need to cover long distances to satisfy the most basic everyday shopping needs.

Big trucks in small places. Photo by the author.

Another issue that needs your support is the elimination of large trucks from areas where people are moving about. There are better alternatives from a livability standpoint to sixteen-ton tractor trailers trying to maneuver on city streets or muscling their way through country villages, as they are in the photo above. The reason that those big trucks are there is that it is less expensive for the companies that are shipping goods to load and operate big trucks fewer times than it is to load many smaller trucks many times. This translates into lower costs for the consumer as well. But we pay other costs as a result, and one of them is that there is now a jousting match taking place on all our roads between cars and people on one side and big trucks on the other, and trucks have the size advantage. Promote and support local legislation to prohibit trucks over three tons on streets where the speed limits are, or should be, 30 mph (50 kph).

As I said at the opening of this chapter, it is not going to be easy to make the tradeoffs between choosing to buy at a small, local shop rather than a large, distant Big Box store. Depending on where you live today, you may not even have this choice. You may not benefit personally from the sacrifices that you make to support small, local shops. You may never experience yourself the rewards from the energy and political capital you invest in bringing these shops back to your community. Nevertheless, you should feel good about the part you are playing in working for positive change by putting the places where we buy things back where they belong: across or down the street, or around the corner from where we live.

Give Commuting a Rest

If you lived here, you would be home now.

THESE WORDS HAVE STUCK IN MY MEMORY since the first time I read them many years ago. They were printed on a sign posted right on Storrow Drive in Boston, Massachusetts at the edge of a high rise, high rent apartment complex. The sign may well still be there. The sign stared commuters in the face every day, taunting them as they struggled with the daily grind of traffic congestion at one of the principal bottlenecks in the city's transportation network. It was here that traffic came to a standstill whenever a drop of rain or a flake of snow fell from the sky. It was here that at least once every few years an errant truck driver would ignore the *Low Bridge* sign and give the top of his trailer a scalping, backing up traffic for a least a mile in both directions.

It is rather easy to make a counter list to this challenging sign:

- If I lived there, home would be an apartment, not my own house.

- If I lived there, I would not have my own lawn and garden outside my door.

- If I lived there, I would not have my own car parked in my own garage next to my own house.

- If I lived there, I would not be able to barbecue in my own back yard with just my own invited guests.

- If I lived there, I would not be able to take a quiet walk in a quiet neighborhood any time of the day or week.

On the other hand, if I lived there, I would not have to worry about cutting the lawn and tending the flower beds, painting and repairing the house, (or paying someone to do these things for me), or perhaps, I would not need to own a car and a garage in which to store the car. Still, there is the barbequing thing.

Commuting is about making tradeoffs, but it is not always a sure bet that living in a city, like the apartments in Boston, brownstones in Manhattan, or terraces in London, will guarantee you a commuting-free life, just as living in a small town or even a suburban setting means that you have to drive for hours to work every day. You may begin with good intentions, to live in a city close to your job. You rent an apartment, buy a condominium or townhouse. Then, suddenly, the city-based company that is only a short walk from your door one day decides to move, lock, stock and barrel, to a rural campus, or it moves your department to a suburban office park. The only way to travel to the new job without spending half of your day in transit is to drive, so if you did not own a car, you purchase one. You would then join the growing number of reverse commuters, those who leave their central city dwellings each day to work in offices outside the city. Then, after hours spent hunting for a parking space each night when you return from the suburbs, and paying countless parking tickets, you decide to rent a garage space. Soon, you are spending so much money on owning a car, and so much time commuting, you decide to move out of the city closer to your job. You find a house in the next village and have a short drive to work each day. Then your company is acquired and your job is eliminated.

You can decide where to live, shop, recreate and where to send your children to school. Deciding where and when to work may be difficult or impossible, and work-related traffic congestion is what gives us the most headaches.

The only way to ensure that your workplace and your place of residence are in close proximity to one another, whether it is in an urban or suburban or rural setting, is to run your own business, and locate it near to, or in, your home. If you are a lawyer, architect, doctor, dentist, barber, or any other type of sole practitioner who can tailor the type of work you do to your immediate surroundings, you may well be able to pull this off.

For the rest of us, the nine-to-fivers—or more likely, the seven-to-seveners—who work for a company that is not down the street or in the same town, we are going to have to develop a strategy that will make commuting more bearable by reducing our daily dose of traffic congestion. A simple and effective way to do this is to reverse the *Principle of Triple Convergence* that I described in an earlier chapter. This holds that traffic will always increase on a new road because people will change route, mode and time of travel to use the new link in the network. They will keep making these changes until the link has been filled to over capacity, and they will stay with the congested route until a new alternative is opened, and when they move to the new route depends on whether they are a *Sheep* or an *Explorer*.

Reversing this means changing all three components, and to do it on a continuous basis. Vary your commute by choosing different routes, modes and times each day.

Change your route. Get off the motorway or the interstate and drive the local roads. It may take you just as long to make the journey, but you will usually be driving at a normal local road pace. Try this: Log on to one of the routing web sites, like MapQuest.com, randmcnally.com, AAAmaps.com or googlemaps.com. Set your home as your origin and your workplace as your destination. Request a route using the default *Fastest Route* setting and compare it to how you normally travel. Is it the same? If not, does it look like it might be an improvement over the way you drive today? Give it a try on a day when you do not have any early morning appointments, or when you can leave before your normal departure time just in case the route turns out to be a nightmare.

Now use the routing web site to produce more alternative routes by selecting Shortest Route, Avoid Tolls Route, Avoid Motorways/Interstates Route. Print out the results and give them a try. You may find that the shortest route is the most pleasant, taking you through parts of your neighborhood you had never before explored. On the other hand, you might find yourself in areas that you would rather have avoided. Use some good judgment, perhaps checking the routes with friends at the office.

Change your time of travel, but not by giving up more and more of your time to both your employer and congestion. Many commuters try to beat traffic by leaving home before congestion starts. The problem is that as more people do this, the length of the congestion period increases toward the earlier times. There is already evidence that this is happening. An article in the Wall Street Journal says that "figures show people are sacrificing sleep for kids, gym, chores and to beat traffic."[87] The article states:

"By a wide variety of indicators, from electricity usage to water consumption, more U.S. households are starting their days before dawn. In the last six years, PJM Interconnection, which supplies electricity to more than 50 million people in 13 states, saw its largest uptick in usage between the hours of 5 a.m. and 7 a.m., while in Atlanta, Southern Co.'s peak winter electricity usage shifted to 7 a.m. from 8 a.m. in 2003.

"For some people, it is simply a matter of trying to beat the traffic. In the last five years, the number of people leaving home between 5 a.m. and 5.30 a.m. increased by 12%, the biggest jump in rush-hour departure times, according to the Census Bureau. That, of course, moves everything else earlier. Quality Care Associates, a child care company that serves several high-powered New York City suburbs, says requests for nannies to start at 7 a.m. are up 5% over the last year."

Since the start of the workday is fairly constant, at between eight and nine in the northern zones, to nine and ten in the southern zones, we gain little by giving up sleep and morning time with our families by trying to get a jump on traffic and leaving before the rooster crows. Eventually, the congestion period catches up with us and we spend as much time in the car as we did when we left earlier, and we end up spending more time at work as well.

Use the congestion curve to your advantage in deciding when to leave the house and when to leave work. The later you leave in the morning, the less time you will spend in traffic. The absolute worst times to leave are between 7.30 a.m. and 8.30 a.m. By leaving at 8.30, your journey is likely to take one-half the time it would if you left between 7.00 and 8.30. So if the drive with traffic takes one hour, and without traffic it takes thirty minutes, if you leave at 7.30 you will arrive at 8.30, but if you leave at 8.30, you will arrive at 9.00 or thereabouts. You could spend the extra half-hour maybe walking to school with your children. The worst time to make the home journey is when most people leave work, between 4.00 p.m. and 7.00 p.m. (16.00 and 19.00 on the graph, which uses the 24-hour continental European method of designating time). Try taking the local roads on the return trip, since they tend to be less trafficked during the evening rush hours, as can be seen on the travel times graph.

Change your mode by taking the train or bus if these possibilities exist for you. Choose a cool, sunny spring or autumn day when there are no pressing early morning meetings. Try to get as much information about the schedule and fares beforehand. Most public transit systems have web sites that you can visit to plan a journey using all available modes. In some places in the world, public transportation is an option of last resort, used only by those who do not have the means to drive their own car. In other places, it is a totally viable option, although one that may make the journey longer or make it less comfortable than driving our own car. If you live in a place where there just is no public transit, or where you would not feel at all comfortable with your fellow riders, maybe it's time to start thinking about what your government is doing with your tax money.

Figure 15: Various daily frequency characteristics for different types of traffic volumes on different types of roads in and around the city of Munich, Germany.[88]

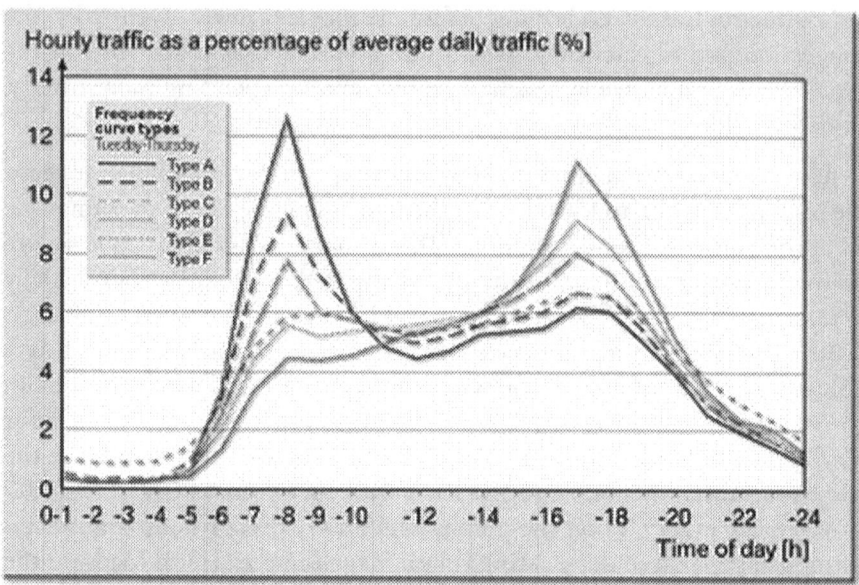

One advantage of public transportation is that you don't have to drive. *"It's such a comfort to take the bus and leave the driving to us."*[89] If you remember that little slogan, you probably remember a time when taking the bus or the train was a pleasurable experience. Riding, as opposed to driving, means that you can do something else while making the journey. Some people sleep; others read a book or the morning newspaper or a magazine; others play video games; others listen to their iPods; still others actually converse with their traveling partners. All of these are good reasons to let someone else do the driving, but some of them are also reasons many commuters give for taking their car. They do not want to listen to other people conversing or hear other people's music. They want to listen to their own music in full surround sound, talk back to the folks calling in to the talk shows, eat their donuts and drink their coffees in the quiet of their own dining car.

Whether or not there was ever a time when the wishes of the majority were respected by all public transportation riders is not terribly relevant. They were respected when I lived in London decades ago; they were respected in New York city on the buses and subways before New York's deterioration in the late 1960s; and they were especially respected in my home town of Scranton, Pennsylvania. There still are in some places where respect for fellow travelers is the rule. On a recent trip to Tokyo, a message over the airport bus requested that people not talk on their mobile phones because it disturbed their fellow passengers, and there is no tolerance—or room—for anti-social behavior on the Tokyo underground. However, today, in most large western cities, everyone does pretty much whatever they want, and don't try telling a teenager to turn down his radio, take her grubby sneakers off the seat, or pick up the candy bar wrapper he just threw on the floor. They can point to adults doing exactly the same things.

Part of the reason for this problem is that "responsible" adults abandoned public transit and thereby removed this balancing mechanism. Before we are going to be able to convince large numbers of would-be public transit riders to get out of their cars and onto the bus or train, we are going to have to address the problem of anti-social behavior. That may involve teaching an entire generation of adults that they need to teach their children the meaning of the term "consideration for others". This will be no small task.

If the public transit option does not exist for you, or if you are too faint-hearted or thin-skinned to try it, and if you cannot tolerate the rigidity of car-pooling, try to set up a private version of public transportation with your employer. This is a new idea that is catching on in parts of the world where park-and-ride facilities have not been popular among politicians and public road authorities. This is the case in many European countries, like Sweden and Great Britain, that have strong public transit histories and very little tolerance for car drivers. The idea is to work with your employer to set up shuttle bus service to ad hoc park-and-ride facilities located at shopping centers.

Finally, use every opportunity to work from home, and mix in vacation days, if you can, to ease the congestion load on the week. This is the ultimate way to beat traffic, by just not getting into it. This may sound like the ultimate cop out, or even a call for early retirement. I would say it's more a matter of pacing yourself to be able to stay the course and finish the race, rather than crashing and burning along the way.

If Not Tolls, What?

THE WORD TOLL HAS MANY MEANINGS, and most of them having negative connotations. As a noun it is a "fixed charge or tax for a privilege, especially for passage across a bridge or along a road". It is also a charge for a service, such as a long distance telephone call. *Toll* as a noun can also mean "an amount or extent of loss or destruction", as in 'The tsunami in the Indian Ocean on December 26, 2004 took a heavy toll on life and property'. As a verb, it means "to sound a bell slowly at regular intervals", or, "to announce an arrival with such sounds". *Toll*, as in "Ask not for whom the bell tolls…" derives from the Middle English *tollen*, which means "to ring an alarm" derived from the Old English word *tyllan* in *fortyllan*, which means "to attract or allure".[90]

The word *toll* as a "tax" comes from the Greek *telos*, tax, through to Old English, *toln*. Tax collector in Greek is *telōnēs*. The Greek for toll booth is *telōneion*, in Latin it is *telōneum*, and in Medieval Latin it is *tolōnium*. Toll as "tax" is translated into French and German as *tribut*, into Italian as *tributo* (He paid the road toll: *Ha pagato il tributo della strada.*).

If a toll is a privilege, someone must own the right to grant the privilege, and everyone else must be in a subservient position to this owner. Perhaps this is one of the reasons that the word has such negative connotations— except for the privileged ones. In earlier times, kings and local lords taxed their subjects to travel on their roads and across their bridges; later, it was a building society or corporation that incurred the debt to pay for the road or bridge; and, today, it is governments who take for themselves the privilege of imposing economic restrictions on movement to achieve a variety of goals, such as to pay for the infrastructure or to channel that movement into collective forms of transportation.

Since there are Latin and Greek words for tolls and toll booths, one might naturally assume that tolls were collected back in ancient times. Yes, it seems that tolls were collected even back then. Rodolfo Lanciani, in his book *Ancient Rome in the Light of Recent Discoveries*[91] wrote: "Travelling on the great consular roads of Italy was always made disagreeable by publicans, or toll and octroi[92] collectors."

Legend has it that the first toll roads in the U.S. were built at the urging of the country's first president, George Washington. The country was in as much debt in relative terms as it is today, and in order to encourage expansion westward, roads were needed.[93] The government could not pay for the new infrastructure, so private companies were encouraged to build the roads in return for having the right (privilege?) to collect tolls. Toll roads in the U.S. are called *turnpikes*, the word probably originating from a turnstile or gate blocking passage until the traveler paid a fee. The first turnpike was the Philadelphia and Lancaster Turnpike built in 1795. Road travel was very much inferior to other modes of transportation back in the pre-car days, and turnpikes had a difficult time competing with railroads and canals. Roads lost their attraction as investments, and toll roads fell out of favor.

After a hiatus of around one hundred fifty years, toll-road building in the U.S. started again toward the middle of the 20th century. The Pennsylvania Turnpike was started in the 1930s and led the new trend. Massachusetts, New Jersey, Maine, New York and other states followed suit with their own turnpikes. Bridges and tunnels in and around these states were also financed by groups who then collected tolls to pay the debt and operations costs.

These were tolls for passage along a road, over a bridge or through a tunnel, and the payment ostensibly was, and continues to be, meant to cover the building and operation of the infrastructure. There are also tolls that are charged for entrance into a city, region or state. From ancient times and well into the 19th century, walls surrounding cities were built to ensure that trade routes passed through the cities' gates. Tolls or customs were levied by the ruler of the city region on people and the goods they carried into the city. The records of customs for he City of London in 1260 are recorded in a volume called *Liber Albus*, from the Corporation of London Records Office:

One penny a dozen of cordwain, one penny the dozen of "godelmynges", one halfpenny the dozen of basan, one halfpenny the pound of silk, one farthing the pound of saffron. Let it be known that this custom is levied only on goods which come from abroad. No customs are due from wax, verdigris, copper, tin, or grey-work, if they do not pass beyond Thames Street towards the north; if they do pass, they shall pay sixpence for a pack, threepence for half a pack, or a penny halfpenny for a quarter of a pack.[94]

In 1856, there were 117 toll gates within a six-mile radius of Charing Cross, the official "center" of London, England. In spite of a ruling in 1825 to reduce or eliminate these toll gates, the number had actually increased from seventy-one in the intervening thirty-one years. Most of the roads had long ago paid off their debts, which was the original reason for the tolls, but their owners continued with extracting payments from travelers. A Parliamentary Committee Report in 1825 motivated the removal of tolls thusly:[95]

> *The very small extent of the trusts, their particular situations, and the necessity of placing the toll-gates of each separate trust within its own little jurisdiction, have had the effect of fixing the toll-gates round London in situations the most inconvenient and vexatious to travellers—an inconvenience which has augmented with the great increase of the suburbs of London, whose intercourse and commerce within the limits of Middlesex has become as upon the streets of provincial towns; hence the frequent payments, stoppages and vexatious delays have become very serious grievances, which still continue to increase, to the great diminution of the value of property.*

An act was past to remedy this situation, but it seems to have had little effect. In a later report, it was noted:

> *Toll-gates have the injurious effect of causing the erection of stables, cowhouses, cattle-sheds, and slaughter houses, within the gates, to save the expense of tolls, thus causing great injury to the health of the inhabitants inside the gates. One of the greatest objections to the system of turnpikes is, that it is a **wasteful and offensive mode of collecting a tax**, entailing a loss or deduction on the gross receipts, difficult to estimate, but of large amount, as the number of men and boys employed is great, and the profits of the contractors or farmers of the tax necessarily very considerable. It is also an impost pressing heavily and unequally on those whose callings require them to use horses and carts, such as the dealers in coals, timber, and provisions.*

> *The toll gates also act most prejudicially to residents beyond the bars, as they discourage public conveyances travelling beyond the turnpikes, and thus, while impeding the development of the omnibus and cab trade, depriving the neighbourhoods outside of the gates of the advantages which those dwelling within them now enjoy.*

Approximately one hundred fifty years later, in February 2003, London re-instituted a tolling system. Instead of bars and turnpikes, it consisted of cameras that photograph the license plates of vehicles entering the so-called *Congestion Zone.* These license plates are optically recognized using sophisticated software, and the resulting digits are compared to a database containing the license plate numbers of individuals who have paid their toll for driving into the zone after 7.00 a.m. and out of the zone before 6.30 p.m. on that particular day. If the license plate owner has not paid his or her toll of £8 prior to 10.00 p.m. on the day of use, or £10 after 10.00 p.m. but prior to midnight of the day, a fine of £100 is levied and a letter is sent to the offending party. It is increased to £150 after 28 days.

Figure 16: The first London Congestion Charging Zone

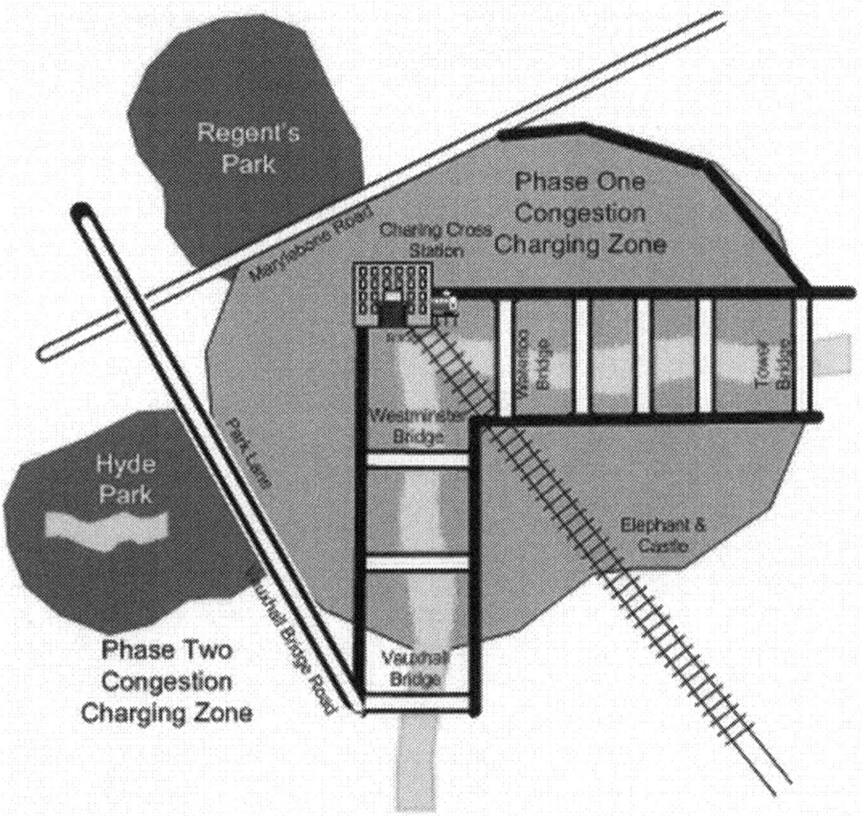

Diagram by the author. Source: Transport for London (2006)

Vehicles that are exempt from the charge are: all two-wheelers; London licensed taxis; London licensed minicabs; certain operational vehicles used by the emergency services, including HM Coastguard and Port Authorities and certain operational vehicles used by the eight London Boroughs, either partly or wholly within the congestion charging zone; the armed forces; Royal Parks Agency; and, breakdown organisations. Residents living inside the zone can apply for a yearly license at a 90% reduction. Obviously, any vehicle that never leaves the zone, or that enters prior to or leaves after the tolls are in force, does not pay the toll.

The infrastructure cost for this system that initially covered eight square miles (nineteen square kilometers) of Greater London's total area (see map above) was £200 million. The annual operating costs are around £115 million. The income generated from the scheme in the first year was £70, almost one-half lower than the projected figure of £130 million. It rose to £97 million in the second and third years after the fee was raised from £5 to £8. This is still approximately £20 million in annual costs that have to be paid for out of other tax revenue sources.

The London scheme was put into place by the citizens of the inner boroughs, led by a powerful mayor, who were fed up with their streets being constantly clogged with traffic. It was a move of desperation. This "use public transportation first" lobby may have been aided by the environmentalists, but the primary goal of providing increased mobility was given much greater weight in the final decision to institute the tolls.

A similar system to London's was put into place in Stockholm on January 3, 2006 on a trial basis. In Sweden, following the previous general election in 2002, the Social Democratic Party, did not have a simple majority. They needed to build a coalition government, and the Environmental Party (a.k.a. The Greens) was the most likely candidate to serve as a partner along with Left Party (formerly the Communist Party). The Environmental Party agreed to build a government with the Social Democrats on the condition that the Stockholm congestion charging trials were approved. There was no disagreement on the need for some measures to control vehicle usage within the city, but there was definitely no unanimity on the method to be used to achieve this control. Setting up a tolls system was not favored by most of the political parties. The Social Democrats, initially opposed to the plan, agreed to implement it in order to be able to form a government.

The Environmental Charge (the opponents called it a Congestion Tax) would be tested for close to seven months and then put to a vote in the City of Stockholm to decide whether it would be scrapped or continued. The cost of this test was approximately €400 million.[96] As opposed to a single flat charge for entering the zone, as in London, there were variable charges based on time of day, and charges were incurred both when entering and exiting the zone. Charges started at 6.30 a.m. and were in force until 6.30 p.m. A toll of €1.00, €1.50 or €2.00 was charged each time the vehicle entered or left the zone up to a maximum of approximately €6.50 for an entire day. The highest charge was for travel in or out between 7:30 and 8:30 a.m., and between 4:00 and 5:30 p.m. Payments could be made either through a direct debit system using a free on-board device placed in the vehicle that was read by the tolling system at the entry/exit points, or retroactively with money payments. The system read the license plates of cars, as in London, and kept track of each car's total toll for the day.

The Stockholm program had three objectives, according to the information distributed to all citizens of the country: to reduce traffic volumes by 10-15% during rush hours; to improve accessibility for buses and cars; and, to improve the environment. After three months, the group responsible for the scheme reported that total road usage inside and around the charging zone dropped by 25%. It seems that 15% of the people who had been using up space on the city's streets were there with no real purpose because they simply disappeared after the congestion charge was instituted. A total of 100,000 "vehicle passages" left the roads, but only 40,000 new riders showed up on the collective transport system. According to reports by the city authorities, they "have not detected any traffic diversion".[97]

The 25% traffic reduction in traffic is similar to the London reduction, and, as in London, it appears to be to the maximum number of cars that can be taken out of the pool of cars before the economic viability of the scheme becomes questionable. In other words, if the price is so high that more than 25% of the drivers avoid the congestion area, the costs of managing the collection of tolls is much larger than the income generated by it, and the costs to businesses and individuals exceeds any benefits from traffic congestion reduction. But Stockholm has said nothing in its goal statement about having a scheme that paid for itself. Greater reductions in the number of cars means higher accessibility for those vehicles that are left on the roads, and also greater reductions in harmful emissions. One important difference with London is that certain types vehicles are exempt on environmental grounds, including those designed to run completely or partially on electricity or a gas other than LPG, or on a fuel blend consisting predominantly of alcohol.[98]

Figure 17: Number of cars passing in or out of Stockholm

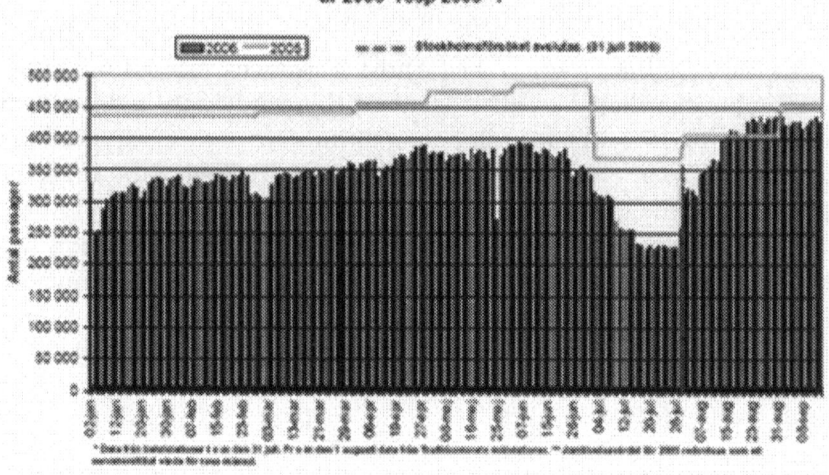

Shortly after the trial closed at the end of July, a report on the local economic effects of the charging scheme that had been commissioned by the authorities was released.[99] Traffic increased 20% the day after the charge was removed (see red line in graph above), it reported, however the time coincided with the return to work after the long Swedish vacation period so it is difficult to judge how much the increase was solely due to the toll being removed. Traffic continued to increase up to and beyond 2005 levels until the end of August, and then declined for reasons that have not been explained..

The test resulted in €14 million less income than projected (€44 million versus a projected €58). This was judged to be the result of a higher than expected number of cars being exempt from the tax, especially those that drove through the city without stopping from an island (Lidingö) that has only one land access point and that is directly into the city. It was agreed that adjustments could be made to improve the economics of the scheme if it is approved for continuation.

Rich men paid the most congestion tax, read the newspaper headlines in the country's major dailies.[100] It was reported that approximately 4% of the private car drivers paid one-third of the total fees. These car drivers were men, high income wage earners, and residents of the inner city. Men in general paid twice as much as women; medium income wage earners were those who reduced their automobile usage the most. In other words, those who continued to drive were those who could afford to do so. They, along with commercial drivers and service vehicles, were the principal beneficiaries of the reduced traffic. Public opinion appeared to move from negative to positive during the trials, and as those in favor gained the upper hand, the Social Democrats shifted from opponents to proponents.

A local referendum was held on September 17, 2006 when the people living in the City of Stockholm voted on whether to keep the system or scrap it. Fourteen of the surrounding twenty-six communities in the County of Stockholm also decided to hold referenda, although their votes were not officially to be considered by the city when deciding whether to re-institute the scheme or look for other alternatives. Of the 620,915 eligible voters in the City of Stockholm, 73.8% showed up to express their opinions. Those in favor of making the "environmental charge" permanent totaled 38.1%; those in favor of dismantling the equipment and not re-instituting the "congestion tax" totaled 33.7%; and the remaining votes cast were either blank or invalid. Not surprisingly, the neighborhoods farthest from the center voted against the scheme, and most of the inner neighborhoods were in favor. The wealthiest neighborhood, Östermalm, which is in the center of the city, voted by a 6% margin against the continuation of the congestion tax. All fourteen of the surrounding communities voted against the tax.

According to the initial ground rules, only the City of Stockholm results would determine the fate of the congestion charging scheme. But ground rules have a way of changing with the winds of political change. On the same day the referenda were being held, the citizens of Sweden were also voting in parliamentary, county and local elections. The Social Democrats, who had ruled for twelve years with the support of other parties, had one of their worst elections in almost a century. An alliance of four center-right parties won the parliamentary elections by a slight margin and formed a new government. It was decided that the fate of the Stockholm congestion charging scheme would be decided not at the local, but at the national level, since, it was argued, the wishes of the surrounding communities must also be considered. Would this mean the death of the scheme? Not necessarily. One of the four new ruling parties, the Center Party, is in favor of the scheme, and there are a few among the ranks of the other parties who appear to be ready to vote against their party line.

It may seem remarkable that many of the strongest proponents of inner city road charging are politicians of the left, who should be, in theory, supporters of egalitarianism. Closing a street for all drivers is fair; leaving it open for all drivers is fair; but closing it for drivers who cannot afford to pay for the privilege discriminates against people with lower incomes. Why, one might ask, is that not self-evident? As a concept, "congestion charging" implies that people who can afford to pay the toll have a greater need to use the road, and therefore a greater right than people who cannot afford it. One highway manager put it bluntly: "Not everyone can travel at the same time. Pricing is the means of rationing."[101]

One explanation for the apparent contradiction is that the environmentalists have been doing an excellent job of using the lowering of toxic emissions argument as a wedge to get congestion charging on the public agendas, as they did in Stockholm to get the congestion charging test implemented. Irrespective of the political, financial or emotional reasons for instituting tolls to travel along roads or into cities, the presence of tolls immediately creates a new order. Removed from rush hour traffic during the first days are all those who cannot afford to pay the tolls, or those, who out of principle, refuse to do so. This has generally been around 15% of travelers. They seek alternatives by either changing time, mode and/or route (Remember Downs's Principle of Triple Convergence back in the second chapter?). That is, they travel out of the rush hour times; they switch to bus, train, foot, cycle or pool car; or they take an alternate route that is not within the toll area. Another group who have been shown to remove themselves permanently from the tolled roads are those who make incidental trips into the toll zone. They represent between 10 and 15% of the travelers. This group stop using the services inside the congestion charging zone, and they do not take public transport into the zone, so they are not recorded in the ranks. However, they do not just disappear. They do their shopping or conduct their business outside the zone, in part helping to fill the parking lots of suburban malls.

In time, a new equilibrium is established. From the baseline established in the wake of congestion charging, traffic volumes <u>increase</u> along with population growth if, and only if, commerce and industry in the area increase as well; or, traffic volumes <u>decrease</u> if the population shrinks and/or if businesses move beyond the tolling zone. The fallacy of congestion charging it that it is a final solution to congestion. If traffic volumes increase over time, rates have to be made substantially higher to create a new shock to the travelers' pocketbooks. If traffic volumes decrease to levels that either do not sustain the costs of collecting the tolls (if the scheme should be self-financing), or if reduced car usage has such a negative effect on the businesses in the tolled areas that they are forced to close, rates will have to be significantly reduced or eliminated altogether, as they were in London in the late 1800s.

If governments of any color on the political spectrum insist on using road charging as a tax revenue option, then it should be a fair tax. Rather than basing the sizes of a toll on the time of day, or—worse—having a flat toll as in London, why not base it on ability to pay and the need to drive? Shouldn't a low income, two-job wage earner who needs to get across town between her third and first shift, have priority to use the roads over a high-income executive? She should pay a lower fee, one that is commensurate with her income and circumstances. Does it really make sense to charge an executive earning a six or seven figure income the same amount to use the roads as a person making the minimum hourly wage? I, for one, do not believe that it does.

How could this differentiation on the basis of fairness be accomplished? Governments have no problems using income tests to gauge what its citizens should pay for certain services, including their state and federal tax burden and how much they are able to pay for their children's higher education. Why not apply the same tests to road usage? If the toll payment technology can match license plates to people who have paid and people who have not, and to track the toll evaders down to their doorsteps in order to deliver a fine, that technology can surely keep tabs on a database that says what fee an individual should pay, and whether the fee has been paid.

Pay-as-you-drive road charging is not the same as congestion charging or road tolls. The intention of pay-as-you-drive is that it serves as a substitute for high fuel taxes, excise taxes and other charges levied on vehicle owners. The idea is that it is revenue-neutral, but that the amount paid is directly related to the actual usage of the infrastructure by the vehicle owner. Those who use the roads most, pay the most. In its simplest form, it takes no more account of the drivers' ability to pay than flat tolls and congestion charging, but it could be adjusted to take income and need into consideration.

Are there alternatives to tolls? There are, and those cities that use them instead of tolls do so for sound reasons. In order to really compare these alternatives to paying directly for driving on a road, over a bridge or through a tunnel, or driving into a district, it is essential to be clear on the fundamental reasons for instituting any form of driving restriction. A list of possible reasons would include the following:

- Provide a source of revenue to pay the debt incurred in the construction of the vehicular infrastructure.

- Provide a source of revenue for repair and maintenance of the infrastructure.

- Provide for economic transfer payments from vehicle users to subsidize collective transport.

- Attempt to provide an economic advantage for the collective transport alternative by making private automobile use prohibitively expensive.

- Reduce the number of private vehicles to increase the average speeds and on-time arrival of buses or surface rail systems.

- Reduce the overall number of vehicles in order to reduce one or more of the following:

  - CO_2 or other emissions;

  - Accidents involving pedestrians or cyclists;

  - The amount of space allocated to vehicular rights of way and parking in order to encourage other land uses; and/or,

  - Noise related to heavy traffic.

What we should notice from this list is that these reasons for setting restrictions is to serve one over-riding goal, which is to improve the quality of life for the people who live, work and recreate in the region served by the transportation network.

"Reducing traffic congestion" as a reason is not valid because it does not include even a hint of the consequences or clearly identify the beneficiaries. Those who <u>do</u> benefit from reduced traffic congestion are those who continue to use the roads. In addition to the bus and taxi drivers, and perhaps residents of the district who might be exempt from paying tolls (one could well ask why they are exempt), those who are not sensitive to the cost (i.e., the wealthy) are the principal beneficiaries. For this latter group, those who can afford to continue to drive, "congestion reduction" becomes a euphemism for getting the economically challenged drivers off the road in order to make more room for themselves, the economically advantaged.

There are other ways to satisfy the above objectives without instituting regressive taxes on all drivers. The following are a few examples.

In Italy, some of the major cities have introduced a driving ban within city zones on private vehicles on odd and even days based on the last digit of the license plate number. On even-numbered days, vehicles with an even number as the last digit may drive, while on odd-numbered days, only vehicles with odd numbers may enter the zone. Not surprisingly, this has led to more license plate theft and an increase in the number of two-car families, but it has also significantly reduced traffic congestion.

One Italian city, Bologna, began in 1972 to introduce restrictions in its historic center. Areas were made pedestrian-only and bus lanes were added to streets, limiting space for private cars and trucks. In 1984, the people of Bologna voted in a local referendum to implement even further restrictions on private automobiles in the center. Access to the center became totally restricted between the hours of 7.00 a.m. and 8.00 p.m., except for certain vehicles, such as hotel guests, taxis, buses, residents and shop owners. Speed limits were reduced to 30 kilometers per hour on all roads and parking spaces were reduced. One of the main objectives achieved was the reduction of carbon monoxide levels by more than 75%. Congestion reduction also improved the efficiency of businesses in the district and increased safety for pedestrians.

The City of Gothenburg, Sweden has attempted to make driving within the old city compound a nightmare without actually closing streets, instituting tolls or charging excessive prices for parking. I can confirm from personal experience that that the city's traffic planners have succeeded. Gothenburg is Sweden's second largest city with approximately half a million residents. It is also home to two of Sweden's vehicle manufacturers, Volvo Cars and Volvo Trucks. It has a old center that has a design reminiscent of Amsterdam, with circumferential canals and radial streets. The city's principal founders in 1640 were Dutch merchants, so the link to Amsterdam is not coincidental. While traffic congestion in Gothenburg was never as severe as in the country's capital, Stockholm, the city's planners decided that they wanted to maintain the quiet, pedestrian-oriented environment that had existed before the advance of car and truck traffic.

In the early 1960s, Gothenburg's central district inside the original fortifications was divided into five traffic zones, and these zones exist to this day. Cars and trucks can drive into each of the zones, but driving between zones is highly restricted. To move between zones, it is necessary to drive out to a low-speed parkway that circles the district and then to drive into the next zone. Traffic was reduced inside the district by almost 50% when the restrictions were introduced, pedestrian and bicycle accidents were reduced by 45%, and buses and trolley significantly improved on-time performance.

Restrictions on the number of parking places, sky-high parking fees for workers and free parking for shoppers have been the most common alternatives to congestion charging in the large northeastern U.S. cities. This approach actually reversed the pre-1980s city planning recommendations for new tower offices in downtown areas to provide a maximum number of parking spaces for employees, usually beneath the building. One example of this is Boston, Massachusetts, one of the oldest cities in the U.S. with a tortuous street pattern in its business and financial district. The city has had bridge and turnpike tolls since cows were grazing on the Boston Common. It has also had one of the most extensive public transportation networks in the country comprising underground, trolley, bus and commuter rail.

Still, by the 1980s, following a construction boom in the city that accompanied the opening of the Quincy Markets in the nation's bicentennial year, 1976, congestion on the clogged arteries threading through the city of Boston seemed to be an intractable problem. Gradually, the parking rates in the central business district were raised from a few dollars per day to over ten times that amount by the mid-1980s. At the same time, large park-and-ride facilities were constructed at the fringes of the city where commuters could park for the entire day for free. Office building continued, but the population of the city began a thirty-year decline starting in 1976. Household size also declined, from 3.3 to 2.4 from 1950 to 1990. Boston has now a very low percentage of the region's population, only 14%, compared to other cities of a similar size, like Baltimore, Cleveland, Milwaukee, Portland and Seattle. The number of jobs in the city actually is higher than the number of residents, 671,000 versus 600,000, with services accounting for half of the total. Boston, San Francisco and Washington, D.C. are the only cities in the U.S. with more jobs than residents. This means that suburban commuting is a major challenge.

Boston has succeeded better than other cities with getting commuters to use public transit, and they have done it without instituting congestion charging schemes. There are tolls on the tunnels and one of the bridges leading into the city; there are tolls on the Massachusetts Turnpike entering the city from the west. However, if one wants to take a little extra time, it is possible to enter Boston without paying one single cent. Almost 40% of Boston workers use public transit to commute, either from their communities on the south shore, north shore or western suburbs, or from the park-and-ride facilities. The inner communities, like Cambridge, Revere, Chelsea, Newton are served by heavy rail, light rail and buses. Massachusetts Bay Transportation Authority ridership has increased by 30% since 1970. Around 15% of residents of Boston walk to work.

Cities and city regions are sensitive organisms. They are born, they grow and prosper. Sometimes they stagnate, decline, fall into decay and die. They have good periods and poor periods, depending on countless factors, including those that can be affected by humans and those that are completely out of human control. When city governments attempt to modify the movement behavior of the people who live in, work in or otherwise use the city, they must be aware that their actions can have unforeseen effects. While the short-term results of instituting congestion charging schemes may be easy to measure in reduced traffic on the streets and reduced emissions in the air, the long-term effects may not be evident until after a long period has passed, after the mayor who pushed for them has gone on to another place. Before rushing into congestion charging, governing bodies should consider all the potential consequences, both positive and negative, and seriously test all of the other available options.

Jakriborg, Sweden, a new town being built between Malmö and Lund on the principles of multi-use zoning, pedestrian- and bicycle-friendly streets, and an architectural style that harmonizes all of life's daily activities. The idea is not to eliminate private cars, but to make them unnecessary. Photo by the author.

Accept Some Friendly Advice

You will not be able to move the mountain of traffic congestion overnight, and the mountain will most probably not disappear for you without a great deal of effort. In other words, beating traffic will take time. You will need some help in the meantime while you are working on getting the kids back on the new sidewalks and bicycle paths your community is building; while you are developing a regimen for yourself and your family of recreating without a short or long drive in the car as a prerequisite; while you are voting for and helping to write a new zoning code that will allow shops to be located within a reasonable distance from your home, and will allow your home to be part of a multiuse settlement pattern like the ones that were prevalent before urban sprawl; and, while you are figuring out ways to make a living without giving up a large portion of your time, and an increasing amount of your income, to a daily commute. Help is on the way, and in some places, it is already available, although it has been a long time coming.

Cars and trucks and buses have been improved significantly during the one hundred years since they first started plying the roads, but, until recently, their method of operation remained unchanged: A single driver controls the speed and direction of the vehicle, communicates with other drivers with signals, and receives instructions on how to use the infrastructure through road signs.

Before 1938, when Buick introduced the first electric turn signal in cars, drivers indicated their intentions to turn with hand signals. An arm straight out the window meant the driver intended to turn left (or right in right-hand drive countries); an arm bent upward meant a turn to the right (or left in right-hand drive countries); and an arm bent downward meant that the driver intended to stop. Lines on the road (solid, long dashed, short dashed, single or double along the curb edge, diamonds at intersections) are still the principal way that drivers are guided in the use of the infrastructure. Simple, two-dimensional signs, like the ones pictured above, plus hundreds of others, tell the driver what he or she can or cannot do. There are other signs that tell the driver what to expect to find around the bend or over the top of the next ridge.

In the world of driving, up until some time around 1996, there was nothing similar to smoke signals, flag semaphores or Morse Code[102] delivered with flashing lights or buzzing sounds, which drivers could use to receive messages from over the horizon warning of an accident, a construction-related detour or closed down traffic lanes, dangerous road conditions, or slow moving traffic caused by just too many cars trying to use the road up ahead. Equivalents of these dynamic signaling systems started appearing as a result of the "convergence of multiple enabling technologies" in *technospeak*, or in plain English, because it was finally possible.

Four things had to happen before drivers could start receiving useful, up-to-the-minute information about traffic conditions in their vehicles, and begin using this information to make better choices about the routes they traveled:

- There had to be ways of collecting traffic data worth sending;

- There had to be ways of processing the collected data into useful information;

- There had to be ways of sending and receiving the information to the vehicle anywhere in the region; and,

- There had to be ways of integrating the information in the vehicle and putting it to use.

Data Collection

The first comprehensive traffic information delivered to moving vehicles was provided by helicopter traffic reports. The helicopter was invented by Leonardo De Vinci in 1480, but the first person to take off vertically and make a free flight entirely without assistance from the ground was a Frenchman named Paul Cornu, who accomplished this feat in 1907. The first practical helicopter came in 1936, the German Focke-Wulf Fw61. Helicopters were first used extensively in military action during the Korean Conflict. In 1950, the Mobile Army Surgical Hospitals (M.A.S.H.) teams in Korea (seen in the popular television series of the same name) were the first to make extensive and practical use of helicopters.

With their bulbous cockpits, helicopters looked like a giant eye in the sky.

The Sikorsky S-55 was the first helicopter to be certified for commercial transport. Traffic reports—Your Eye in the Sky—started in earnest in the mid-1970s, although the first traffic report from a vertical flyer reportedly was made twenty years earlier in Richmond, VA. Helicopters are still one of the most effective ways of communicating traffic conditions to the individual driver because they can see what the driver cannot, the road ahead. However, helicopter broadcasts continue to have one major drawback. They always seem to deliver their reports when drivers are already stuck in traffic, rather than when the incident that would eventually cause the log jam had just happened so that drivers could avoid the area. And, since the helicopter cannot be everywhere at once, some events are simply missed. The cost of running the broadcasts continuously, or of having a large fleet of helicopters hovering constantly over the city, are apparently too high to even consider. A more constant and reliable source of traffic information was needed.

Public road authorities began to monitor numbers and speeds of vehicles on certain stretches of roads in order to better manage these flows, as well as to gather data for future road maintenance and building projects. This work began in earnest in the 1950s in the U.S. and a bit later in Europe and Japan. If a road has a certain capacity, and it can be shown conclusively that this capacity is regularly exceeded, that should be sufficient justification to obtain the funds to expand the road by adding lanes, or to build a new road that would divert cars from the congested one. This is what public road authorities did, they built new roads, maintained and added more lanes to existing roads.

The need to measure vehicular traffic led to the wide-scale introduction of traffic flow measuring instruments in most city regions. There are some very sophisticated technologies developed for measuring traffic flow. The most common, which is inductive loops, is based on sensors embedded into the road that measure the time it takes for a vehicle to pass over the sensors, and at the same time counting the number of vehicles moving along the road. The sensors are connected back to a central data processing center through a communications network, and the data is processed into average speed of travel.

Another technology that is used for measuring traffic flow is optical sensors, which is based on video cameras. One set of cameras constantly photographs the license plates of cars passing, and a second set further along the road takes another photo. Sophisticated optical character recognition software "reads" the licenses and records the time that the vehicle has passed each camera. The distance between the cameras is fixed, so the only variable is the time it takes for the vehicle to move between the two locations. Distance divided by time equals speed.

Figure 18: Traffic flow measuring with optical sensors

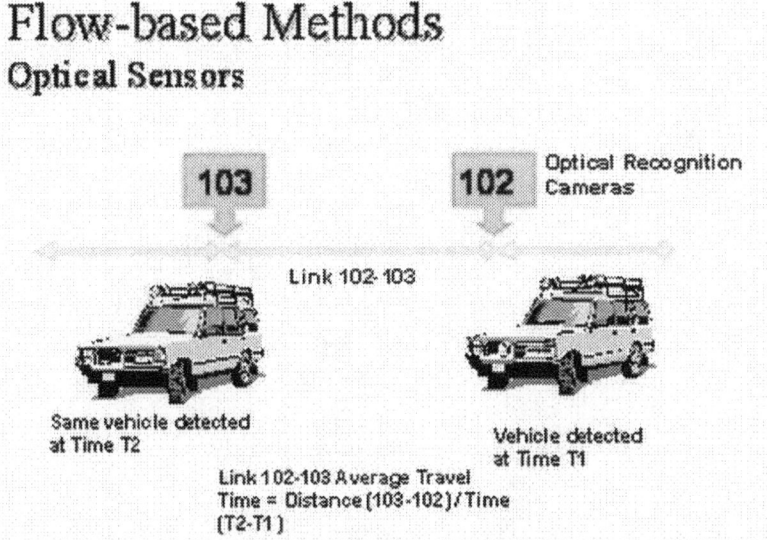

Other common methods of collecting traffic data are monitoring police reports, assembling road authority bulletins, or encouraging people to call in observed traffic problems by making them "official" traffic reporters, the *Jam Busters*. Traffic data suppliers are constantly increasing their coverage of roads by installing new traffic flow sensors, adding more traffic reporters, and improving cooperation with police and emergency authorities so that they provide faster and more accurate input. It turns out that these devices are also useful for collecting information that is helpful to drivers, such as where the roads are congested and where traffic is flowing along at a reasonably normal pace so the congested roads can be avoided and the freely flowing roads can be used instead.

But sensors, whether they are embedded in or fixed above the road, are expensive to install and difficult and expensive to maintain. Cameras are often vandalized because people confuse them with speed cameras. Due to their high cost, they are impractical to place on secondary and minor roads. The Holy Grail of traffic data collection is the equivalent of sensors on all roads.

So-called "floating car data systems" offer a promising alternative. Floating car data systems have nothing to do with cars physically floating over the roadway. It is a rather ingenious term, invented, no doubt, by one of the transportation system intelligencia in Europe or North America working on ways to reduce congestion without building new roads. The word **float** is Old English, *flotian*, meaning "body of water". There are many definitions of float, all more or less connected to doing something on or near water. The two definitions that best apply to the floating car application are: 1) to move with a moving liquid: to drift; ; and, 2) to pass from person to person, as in "The rumor floated through the town". The word **float** is related to the word **fleet**, which is "a group of vessels or vehicles moving or working together".

The Floating Car Data (FCD) technique is based on cars being able to travel anywhere and deliver data from wherever they are. The cars are "floaters", and are equipped with positioning technology and wireless data communications. They send their speed of travel and positions to a central server. The server collects travel time information from vehicles and matches the locations to digital maps. The result is a map of average speeds along segments of the road network.

Another term that is used for the same technique is Probe Data. **Probe** in the verb form means "to search into or examine thoroughly. As a noun, a **probe** is "an investigation", or it is "a vessel carrying scientific instruments to record or report back information about space or other planets—or even what is behind enemy lines". **Probe** is derived from the Latin, *proba*, "a proof", or *probare*, "prove".[103]

Floating car data, or, better yet, floating vehicle data, is a more appropriate term for describing the technique, and that probe is a good name for one of the vehicles working in the fleet.

Information Processing

The key to using all of this traffic data is to put the data into a form that is useful for helping drivers make pre-trip and on-trip decisions about what is the best alternative route to take to complete their journeys in the fastest possible times. Starting in the late 1980s, groups in Europe and Japan, operating independently, developed methods for coding traffic data collected using the techniques described above. In Europe, they chose to work backwards from the end use applications in order to design a coding system that could meet their usage requirements. The principal requirement was to convert a digitally coded text message into an audible message that would be played through the vehicle's radio, and to be able to broadcast a single message in multiple languages.

The system that was developed in Europe was built on top of the Radio Data System (RDS). Radio Data System has been operational in Europe since 1987. It was originally developed by the European Broadcasting Union, and uses a technique of adding data to a sub-carrier on an existing stereo transmission in such a way that the data is carried inaudibly. RDS supports program-related features, such as tuning to alternative frequencies to obtain the best signal for a channel, and non-program-related features, such as broadcasting digitally coded data on the Traffic Message Channel (TMC) that is reserved within the RDS specification for traffic messaging. RDS-TMC provides a system for collating real-time traffic-related data in a digital coded form, and broadcasting it over standard FM radio channels.

Free traffic information is broadcast via RDS-TMC in a number of European countries, including Germany, Italy, The Netherlands, Sweden, Finland, Denmark, Spain, Switzerland, Austria and France. The UK has no free RDS-TMC services. Private traffic collecting companies, *ITIS* and *Trafficmaster*, have licenses for RDS-TMC broadcasting in the UK. They charge fees to hardware manufacturers or vehicle manufacturers who incorporate methods for reading and decrypting their broadcast traffic data.

The European system, has now spread to other parts of the world, including North America and Australia, and its proponents are trying to make it a standard in China. There they have competition from the Japanese system, which is called VICS.

VICS, for Vehicle Information and Communication System, employs information collection techniques that are similar to those used in RDS-TMC. The traffic data is processed using a national map database originally created by the Japan Digital Road Map Association (JDRMA). All of the navigation systems in the country use the coding scheme created by JDRMA, and the Japan Road Traffic Information Center applies the traffic information to this database and delivers the information to vehicles equipped with VICS-enabled systems via radio beacons, infrared beacons and FM multiplex broadcast techniques.

Figure 19: Traffic delivery in Japan

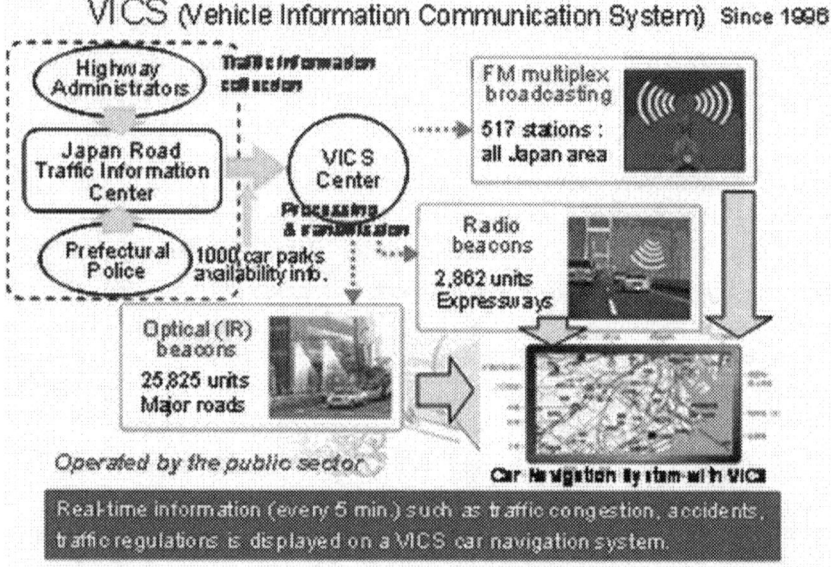

While Europe and Japan were building robust traffic information collection and delivery systems in the 1990s, putting traffic information into a useful form for U.S. drivers proved to be difficult. The main reason is that each state and each major metropolitan area within the states have different ways of collecting traffic data, and, although many private companies tried to create a single service based on these diverse sources, most failed. While Europe was building services based on its RDS-TMC standard, and all of Japan's industry and government were behind VICS, there was nothing that was emerging from the U.S. except more diversity. Another reason is that there is no equivalent to national radio channels in the U.S., like BBC in the U.K., that could serve as bearers for an entire country. Even National Public Radio (NPR), like my old favorite WGBH in Boston, is broadcast by local public radio channels. Driving from Boston to Washington, D.C., you might be able to find NPR channels for most of the distance, but they will be different channels with different ownership, and they will not be airing the same programs at the same time. So you might lose WGBH broadcasting *Click and Clack the Tappet Brothers* in Toll House, MA—just as the Tom and Ray begin to hit their stride—and pick up another NPR channel from Hartford, CT in the middle of its bi-annual fund drive.

Putting together a group of channels in the U.S., one for each market, that could support the broadcasting of traffic data proved to be too daunting a task. Then, starting in early 2003, the two principal map database providers in North America, Navteq and Tele Atlas, cooperated to build a coded road network that could be used for traffic reporting, and to deliver consistently coded traffic information on a country-wide basis. They adopted the RDS-TMC coding used in Europe. To deliver the information to vehicles, they engaged the fledgling satellite radio network operators, **XM Satellite Radio** and **Sirius Satellite Radio** as a substitute for FM broadcasting. By the end of 2005, there were nine million subscribers to satellite radio, and traffic information was being broadcast on both networks.

Information Sending and Receiving

The car radio was an invention that had potential for aiding the driver by getting information into the vehicle and potentially delivering it to the driver, as it is doing today with FM and digital audio broadcasting, both satellite and terrestrial. Paul Galvin, president of Galvin Manufacturing Corporation, apparently invented the first car radio in 1929. He coined the name "Motorola" for the company's new products by combining the notions of motion and radio. The first car radios were not sold by the car manufacturers, but were purchased separately. It took over thirty years from the time cars were first wired for radio in 1930 until the medium was used for informing drivers about traffic conditions with helicopter broadcasts, and another forty years before radio was used as a data receiver for traffic information.

Traffic via RDS-TMC became standard in all navigation systems sold in Europe during the first half of the 2000s, and it is becoming a de facto standard in satellite digital audio broadcast systems, like XM-Radio and Sirius Satellite Radio in North America. Broadcast radio, both analog FM and digital, is an excellent data reception device, but it has one major drawback: it is only one-directional. It is great for sending data from a single point to multiple receivers, but it doesn't work the other way around. Two-way communication is useful when you want to ask for something in particular. For that, another form of radio is needed: the *telephone* (Greek for "distant voice").

Telephones have become devices that are equally adept at sending and receiving data as they are at sending and receiving voice calls. And yet, since telephone technologies in all the major markets are different—why would anyone expect the phones to be standard when we still have metric and imperial measures, different television technology, different electrical systems, different driving sides of the road, and countless other differences?— solutions that are developed in one market do not necessarily work in the others. (See sidebar *Wireless Telephone Systems in the World.*)

Figure 20: Illustration of car kit with phone holder and Bluetooth GPS device on top of instrument panel.

Wireless Telephone Systems in the World

<u>GSM</u>: Global System of Mobile Communication. GSM is global, operating in over 200 countries (not Japan) - 900/1800 Mhz, 1900 Mhz in U.S.

<u>GSM/SMS</u>: Global System of Mobile Communication/Short Message Service. Provides for 160-character messages between GSM handsets or between a GSM handset and an information source. Messages are entered via the handset's keypad, or via add-on keyboards, such as Tegic's T9 or Ericsson's Chatboard.

<u>CDPD</u>: Cellular Digital Packet Data. A method for sending packet data over analog networks. Requires specific network equipment, that now covers 85% of North America to support wireless Internet connections

GPRS: General Packet Radio Service for GSM, enables more efficient use of radio resources leading to increased capacity and higher speed data services

<u>UMTS</u>: •Universal Mobile Telecommunication System - Also known as 3G (for 3rd Generation) - Officially IMT-2000

–WCDMA - Wide-band CDMA. Supported by GSM countries and Japan

–CDMA-One - Developed by Qualcomm

Information Integration and Usage

It's one thing to have good traffic information; and even to be able to make specific requests about what you want, where and when you want it. It's quite another to actually be able to do something useful with it, especially in unfamiliar territory. Except for providing you with peace of mind to know that there is a two-mile queue waiting for you up ahead, and that it is due to a collision between a truck and a passenger car, this knowledge does not help you get to where you want to go any faster. What would be useful is if another route was proposed that could get you around the jam safely and quickly. This is where in-vehicle navigation combined with real-time traffic information comes into play.

In-vehicle navigation is a general term referring to any one of a number of wayfinding methods. People driving cars navigating to their destinations using visual landmarks as waypoints, following street signs and route number signs, referring to maps, and asking people in the vicinity for directions, are all wayfinding methods. Paper maps and atlases are still the most common tools used by people all over the world to find their way in unfamiliar places.

Map reading is not a skill that everyone learns, or that everyone can master. It requires a significant amount of mental gymnastics to relate a three-dimensional world to a two-dimensional representation of that world depicting a view from several thousand meters up in the air. It is difficult enough to use a map while on foot, but it is a dangerous practice for a driver riding alone to try to use a map while the car is in motion. In 1911, the American Automobile Association started publishing road maps. They were not standard topographic maps, but consisted of route descriptions, describing where an automobile and its passengers could travel in relative safety. This concept was eventually developed into the AAA TripTik.

Route numbers, road names and road signs became essential for wayfinding as the number of car-usable roads grew. Before route numbers, which were first used in the UK in 1921 and in the US in 1926, major roads were usually named after the principal destination, such as the Boston Post Road, or a geographic location, like the Blue Ridge Parkway, Mohawk Trail, Dixie Highway, Pikes Peak Ocean Highway, or after a road building authority, like the Pennsylvania Turnpike. [104] It is interesting to note that there are places in the world where many streets do not have names. Japan is one example. The major highways have numbers, and the largest arterial roads may have names, but the roads along which most people live are not named. Address coding is quite different in Japan. There, a house has a number within a numbered sub-block; a sub-block is within a named block; a block is within a named ward; and a ward is within a municipality. Whether using maps or a navigation system, wayfinding in Japan involves a nested search: First the city, then the ward and so on down to the house. It also means that navigation can occur in more manageable segments than in the street-oriented western cities.

The automated systems of today have taken some of these older techniques and incorporated them into their solution packages. There are different opinions on what constitutes a navigation system, and who was first to introduce one. The U.S company Etak presented their *TravelPilot* in the early 1980s. *TravelPilot* displayed the position of the vehicle on a map, along with the destination. It did not provide turn-by-turn instructions, but displayed the position of the moving vehicle on a monocolor map. One unique feature of this early system was that the heading of the vehicle on the map screen was always straight ahead, and the map rotated, like when a person reading a map rotates it to align with the street pattern. This was called "heading-up" display, not to be confused with a "heads-up" display, which creates an image in front of the driver's eyes without a physical display, like a hologram.

Japanese manufacturers introduced map display systems even earlier, also without turn-by-turn instructions, using waypoint routing instead. Think of waypoints like the flag buoys in a sailing race. Reaching the finish line (destination) means passing all of the buoys (waypoints). The first turn-by-turn instruction systems were developed in the late 1980s by two large consumer products companies, the German company Bosch Blaupunkt and a division of the Dutch company Philips. Bosch licensed the Etak technology, and still calls their system *TravelPilot*. The first commercial turn-by-turn systems in North America and Europe were introduced in the mid-1990s: the Magellan NeverLost system (originally developed by Zexel), the Bosch TravelPilot (originally based on Etak technology) and the Philips Carin System.

There were four important converging technologies that allowed in-vehicle navigation to reach the consumer market in the mid-1990s and eventually produce systems like the *Volvo RTI System* shown below:

- CD-ROM for data storage;
- Small format, high resolution color display screens;
- The Global Positioning System (GPS); and,
- Digital navigable maps.

Figure 21: Volvo Cars Road and Traffic Information.(RTI) Navigation System.

CD-ROM is an abbreviation for "Compact Disc Read-Only Memory". A CD-ROM is a flat, metalized plastic disc with digital information encoded on it in a spiral from the center to the outside edge.[105] It is a non-volatile optical data storage medium using the same physical format as an audio CD. It is readable by a computer with a CD-ROM drive, and conforms to the CD-ROM Yellow Book standard established in 1985 by Sony and Philips B.V.

The standard CD-ROM can hold 650-700 megabytes of data. It is popular for distributing software (remember when software was delivered on floppy disks?), especially multimedia applications, and large databases, like the geographic databases required for navigation applications. One CD-ROM can store a navigation database covering New England or one-half of Germany. To put that into a more familiar context, that is the equivalent of around one thousand novels, assuming that an average novel contains 60,000 words and each word length is an average of ten letters. Each letter or character occupies one byte, so an average novel contains around 600,000 bytes. One CD is 650,000,000 bytes.

CD-ROM

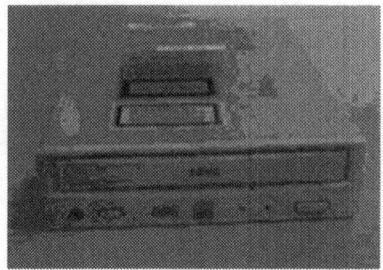

CD-ROM Drive

CD-ROM drives are rated with a speed factor relative to music CDs: 1x or 1-speed which gives a data transfer rate of 150 kilobytes per second in the most common data format. For example, an 8x CD-ROM data transfer rate would be 1.2 megabytes per second. Above 12x speed, there are problems with vibration and heat.

In comparison, DVDs (*Digital Versatile Disc* or *Digital Video Disc*) is an optical disc storage media format that can be used for data storage, including movies with high video and sound quality. DVDs resemble compact discs since their physical dimensions are the same (12 cm), but they are encoded in a different format and at a much higher density. They typically contain 4.7 gigabytes (one gigabyte is equal to one thousand megabytes) of data or more, and have replaced CDs in most navigation systems. Dual layer DVD+R discs contain 8.5GB of data.

Display screens are what make navigation systems visible and real. Everything else is smoke and mirrors. When we say display screens, we are referring to liquid crystal displays, or LCDs for short. These are thin, flat display devices made up of arrays of "pixels" in front of a light source or reflector. The technology traces its history back to 1911, when Charles Mauguin described the structure and properties of liquid crystals. In 1936, the Marconi Wireless Telegraph Company patented the first practical application of the technology. The first operational LCD was introduced in 1968 by a group at RCA in the U.S.A., and by 1971, the technology that is in use today was commercialized by its discoverers at Kent State University in Ohio, U.S.A. What James Fergason discovered at Kent State was the twisted nematic field effect in liquid crystals. By altering the twisting of a liquid crystal, the amount of light moving through it, and the subsequent "color" displayed, could be controlled.

We hear and read the term *Satellite Navigation*, and maybe we believe that satellites are literally guiding our vehicle from where we are to where we want to be. In the December 5, 2005 issue of the magazine *Newsweek*, in The Technologist section (p. 12), a reader asked The Technologist: "I'm finally ready to buy a Global Positioning System (GPS) for my car. Which features are the most important for me to consider?" What the technologist should have answered was: "Do you mean that you will buy a navigation system? A GPS device won't do you much good on its own unless you just want to see your latitude/longitude." The Technologist, first by printing the question, then by assuming that the person simply got his terms wrong and not correcting him (the writer was male), gave the error credibility and proliferated it.

The satellite device provides the positioning part of the navigation function. While this is only one of many functions that need to be performed for route guidance, it is an important one, and understanding how a Global Navigation Satellite System (GNSS)[106] like GPS actually works can provide insights to many of the wonders of intelligent transport systems.

Figure 22: A model for a new type of route guidance system

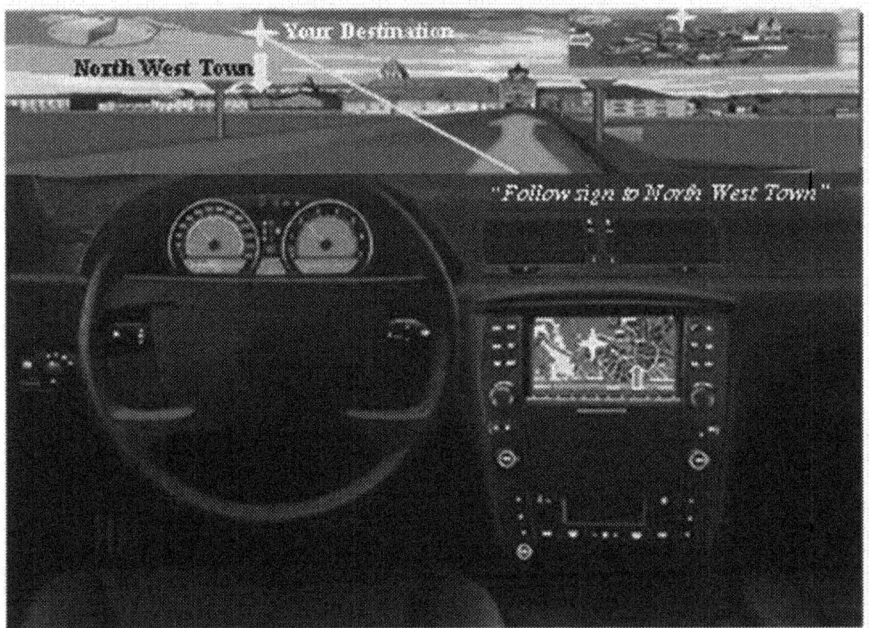

I have been working on a navigation solution that places a virtual star in the windshield—which is in its entirety a heads-up display screen—directly over the destination. The star shines a light beam down on the destination, moving closer to the ground as the driver approaches it. Instructions direct the driver to follow signs and landmarks, and an orientation map is provided in a secondary screen only for reference and to show specific points of interest requested by the driver. Photo collage by the author.

The GPS constellation of satellites was put up in sky by the U.S. Department of Defense, with work starting in 1973. Their original objective for the satellite system was to provide the U.S. military in the first instance, and its allies in the second instance, with a method of identifying the location of any place on, above or below the surface of the earth for any purpose they believed was important for their respective country's security. The system was called NAVSTAR GPS. The first satellite was put into orbit in February 1978. the first civilian use of the system came in 1984, but full-fledged use began in July 1995 when the full complement of twenty-four primary and one reserve were in orbit. Three more spares have since been added.

Initially the military were able to receive signals with a precision of 18 meters, while civilian signals were deliberately downgraded to 100 meters positional accuracy. President Clinton, through an executive order, eliminated the so-called Selective Availability in May 2000.

The satellites orbit the earth in six fixed paths and at constant altitudes of 20,000 kilometers (12,000 miles). The satellites have an orbit time of 11 hours 57 minutes and 58.3 seconds. On board each satellite is an atomic clock that has a precision stability of one second in 150,000 years. There is a computer with sufficient memory and storage for it to receive signals from stations on earth and store information contained in these messages. It also contains a radio transmission unit that allows the satellite to send messages once every second toward the earth, and a power source to keep it all running.

The satellites do not send their positions in their messages. They send the time on their atomic clocks, along with the latest information that was beamed up to them. Every satellite has its own specific flight path that has been precisely plotted on a great circle (actually, an ellipsoid) on the earth. The one-second intervals at which each satellite sends a message correspond to an exact position along this ellipsoid, with the position given in latitude and longitude. The latitude and longitude values for each second for each satellite are calculated from a reference value stored in a table, and this table is stored in the GPS receiver that is part of the navigation system. Why not store the tables for each satellite on the respective satellites and just send the coordinates? In the first place, the time would still need to be sent, for reasons I will explain shortly. It also turns out that it is much more efficient to manage the location information from the ground, rather than trying to keep the satellites constantly updated.

So, each satellite's position can be accurately tracked, but no satellite knows where your receiver is. This must be calculated. In order to complete the calculation, that is, to know the latitude and longitude of the receiver, signals from at least three satellites are needed. The GPS receiver processes a satellite message with the time the message was sent. It looks for the time in its table for the particular satellite, finds the latitude and longitude and notes its location. It then compares the time it received the message on its own clock and obtains a time differential. This time differential t multiplied by the speed of light (c)[107]—that's how fast the message is sent along a radio signal—results in a vector that has a length of d. When one end of this vector is placed on a point perpendicular to the earth's surface directly above the satellite's plotted latitude and longitude, and at an altitude equal to the height of the satellite, which is constant, and the other end is placed on the earth's surface, the vector forms the hypotenuse of a triangle with the base of the triangle being the distance from the satellite's plotted latitude and longitude to the receiver. But with only one satellite, we do not know exactly where the vector touches the earth. The receiver can be anywhere on a circle with a radius equal to the length of the base.

When the process is repeated with a second satellite, we have narrowed down the location to two positions, where the two circles intersect one another. This is an improvement over anywhere on a circle, but it is still not good enough. With a third satellite, we have zeroed in on a position that is a good approximation of our receiver's location. The more satellites we have to work with—up to 12 is possible at any one time—the better the result. When a fourth satellite is available, the GPS unit can also calculate the units distance above sea level.

Now think about doing all these calculations with three-to-twelve satellites every second and delivering a location that is within ten meters of actual ground truth, and you begin to appreciate just how impressive these little devices are. They are even better. The data that are sent up to the satellites that I mentioned earlier, they are corrections to each satellite's latitude and longitude. These corrections are calculated on a regular basis and form supplemental information to the data stored in the tables. The reason that this supplemental information is necessary is that the earth moves. For example, following the tsunami in the Indian Ocean on 26 December 2004, it has been reported that some of the smaller islands southwest of Sumatra may have moved southwest up to 20 m (66 ft), and the northern tip of Sumatra could have been displaced by as much as 36 m (118 ft) southwest. Such occurrences are not isolated. To include such information, it would be an impractical exercise to update all the tables in all the GPS devices in use. Therefore, periodic adjustments are sent up to the satellites and then beamed back in the messages. These adjustments are factored into each receiver's calculations.

Figure 23: U.S. Department of Defense Global Positioning System

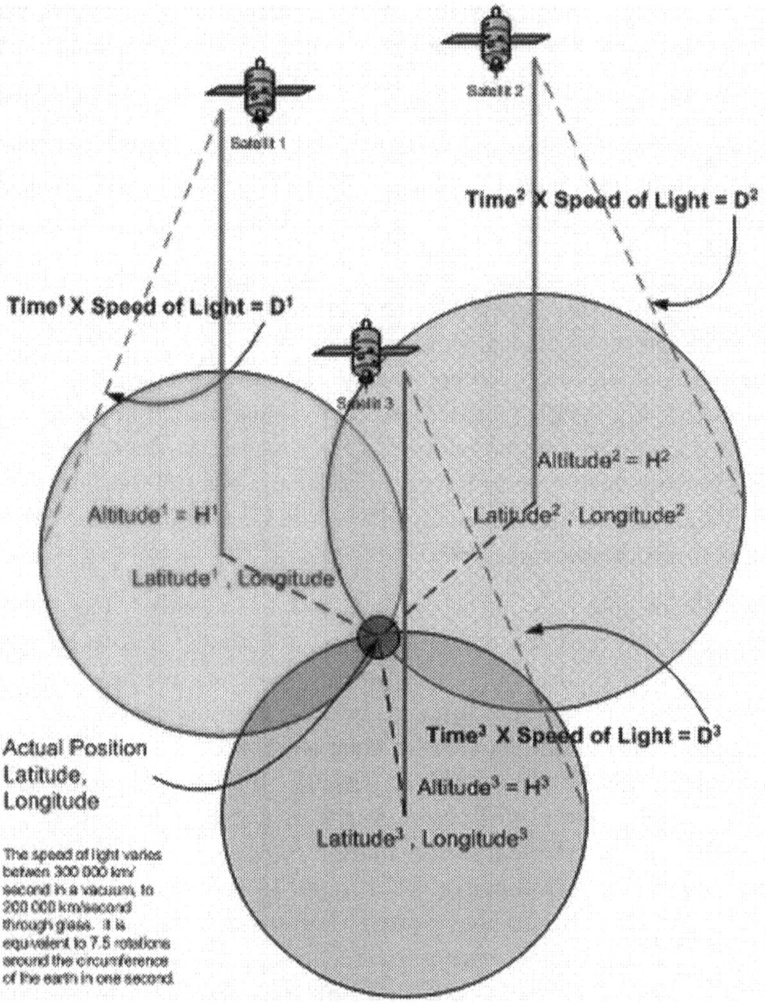

$Time^2$ X Speed of Light = D^2

$Time^1$ X Speed of Light = D^1

$Altitude^2 = H^2$

$Latitude^2$, $Longitude^2$

$Altitude^1 = H^1$

$Latitude^1$, $Longitude^1$

$Time^3$ X Speed of Light = D^3

Actual Position
Latitude,
Longitude

$Altitude^3 = H^3$

$Latitude^3$, $Longitude^3$

The speed of light varies between 300 000 km/ second in a vacuum, to 200 000 km/second through glass. It is equivalent to 7.5 rotations around the circumference of the earth in one second.

Source: Diagram by the author

One last twist. If a GPS receiver had to have an atomic clock, rather than a more normal timing device, no one except the military would be able to afford the technology. The clocks in the GPS devices are simple digital quartz clocks found in everyday electronic products. The device manufacturers distinguish themselves by their skill at making their clocks think and act like they are super expensive timepieces.

In addition to calculating the geographic position of the device, a GPS unit can also provide speed and direction of travel. Speed is simply distance divided by time. The distance is derived by subtracting the most recent position from the last position and then converting degrees, minutes and seconds to meters or feet. The time is provided directly. Direction, or bearing, is derived by measuring the angle of change from the one-second vectors. All of this information is useful for navigating, but it is not sufficient. Plotting the position on a digital map is useful for orientation, but it is not enough to provide the turn-by-turn directions that navigation systems offer today: "Keep to the right to exit the motorway in three kilometers." or "Take the third exit in the roundabout onto Mulberry Street."

To perform this magic, to put all of this information into a useful package for the driver, highly accurate and attribute rich digital map data is required. It turns out that the satellite positioning part of navigation is relatively minor.[108] It replaces the initial calibration of the system at a known intersection. After the GPS fix is obtained, and the system knows approximately where it is, the systems rely on matching the position using the inertial components, bearing and distance moved, to a digital map. When the vehicle is started and begins driving, the GPS positions create a set of vectors that can be matched to the digital map. If a match is found, the system places the vehicle on the road and follows the road using distance and bearing, all the while doing sanity checks in the background with the GPS signals. Every time the vehicle makes a turn, the system re-calibrates itself, matching the calculated position with the actual.

Figure 24: Map matching the position of the vehicle with the vector map data stored on the navigation system.

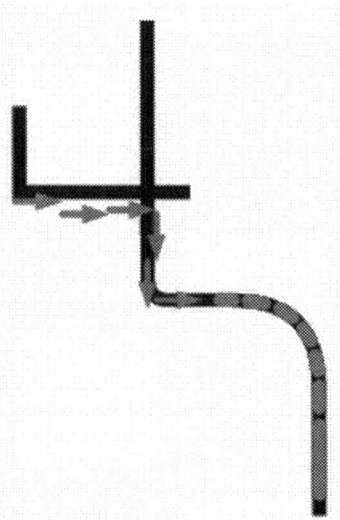

When a vehicle starts moving, vectors are created with the GPS positions. Once the vectors approximate a road network, navigation software matches the vehicle's movement with the map geometry using bearing from the gyroscope, and distance from the odometer.

The national land surveys, census authorities and various federal, state and municipal agencies had begun to assemble digital representations of their road systems in the 1970s, but this data was organized in a way that made it useful for paper map production, but not for use in navigation systems. Private map publishers were making maps with computer techniques as early as 1978, but none had invested in building navigation quality datasets.[109]

New companies were formed to create the necessary databases. One Japanese and two American companies share the credit for having created the first truly navigable map databases. As is usual with inventions, necessity was its mother. Etak in Palo Alto, CA, created a navigation apparatus mentioned earlier. It was not really a turn-by-turn device, but something that was its forerunner. They started building map databases for their system in the early 1980s. Navigation Technologies, which today is called Navteq, grew out of a company just up the road from Etak in Sunnyvale, CA called *Karlin and Collins* that had created a free-standing, vending-machine-type device that would dispense turn-by-turn directions in text form. They, too, found that their device needed data that did not exist anywhere on the planet. The Japan Digital Road Map Association started building a digital map of Japan around the same time, basing it on the paper map sheets created by the country's land survey.

All of these pieces finally began to fall into place around 1995 in Europe, earlier in Japan, later in North America. Mercedes, BMW, Audi and other high-end car manufacturers began offering navigation systems as an option in Europe around the middle of the 1990s. Within a few years, all brands had optional navigation systems. These systems were totally self-contained, with the necessary data and software bundled with the hardware required for operation. Everything was on-board. Navteq, one of major suppliers of navigation databases, even made the phrase its trademark: *Navteq on board*.

By the early 2000s, most navigation systems sold in Europe were able to receive traffic information and use this information to generate new routes that would avoid the reported traffic congestion. It took another five years before North America had similar advantages. Portable devices, like the TomTom *Go*, moved in-vehicle navigation from the luxury category to an every-person's tool. By 2004, portable devices were outselling integrated systems, and by the end of 2005, seven million of them were sold in Europe.

Figure 25: Sales of different types of navigation systems

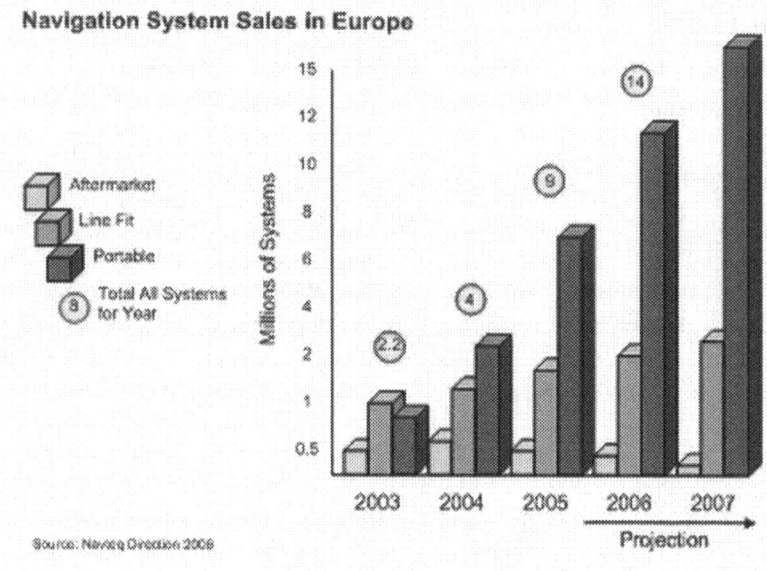

Many navigation systems on the market in Europe in 2005, like the Blaupunkt TravelPilot pictured at right, were equipped with RDS-TMC receivers, allowing them to receive traffic information from public authorities or private traffic providers directly via FM broadcasts. In the U.S., satellite radios started to be used for delivering real-time traffic in 2005.

Figure 26: Bosch Blaupunkt TravelPilot Navigation System.

Making autonomous systems dynamic is one way to improve the performance of navigation systems, and as we said earlier, this is exactly what was done. The downsides of this approach are the costs and complexities they add to the systems, and all of the supporting processes required to keep these systems in operation. The main problem is that the systems are only as good as the on-board data. To make this data truly usable, it needs constant updating, including the coding of new locations for which traffic information can be distributed. The navigation databases must then be delivered to each navigation system supplier for conversion to their particular physical storage format on a CD or DVD or other storage medium, and eventually supplied to the navigation systems' owners.

One alternative to making autonomous systems dynamic is to move the entire data assembly and processing task to centralized servers. In other words, eliminate the CD, DVD or hard disk storage on the device and provide drivers with routing information from the servers that already accounts for the current traffic conditions. This is the *off-board navigation* approach. The data is delivered over the telecommunications data channel, like GPRS for GSM phones.

Most off-board navigation systems attempt to transfer the entire set of turn-by-turn route instructions, and as many digital maps as on-board memory allows. In addition, they generally supply a corridor of data along the route for use in the event a short detour or a wrong turn is made and the user must be re-routed back to the original route.

Figure 27: Appello Wisepilot Smart Phone
Off Platforms

When navigation systems were first introduced commercially in the mid-1990s, there was only one kind, and that was a system built into a vehicle, either completely integrated in the factory or installed by the vehicle or system dealer. Today, there is a wide variety of system types, from the vehicle-integrated to the totally portable, from radio-based to phone-based. The number of DVD- and hard disk-based on-board and personal navigation device solutions are growing quickly. Off-board solutions are at an early stage, and are increasing in number, but slowly. The relative share of PDA-based solutions are, for the time being, diminishing. The PND (personal navigation device) segment started to sail off the charts in 2005. In that year, PNDs outsold integrated navigation systems four-to-one. (See charts on the next page.)

Figure 28: Different types of navigation systems

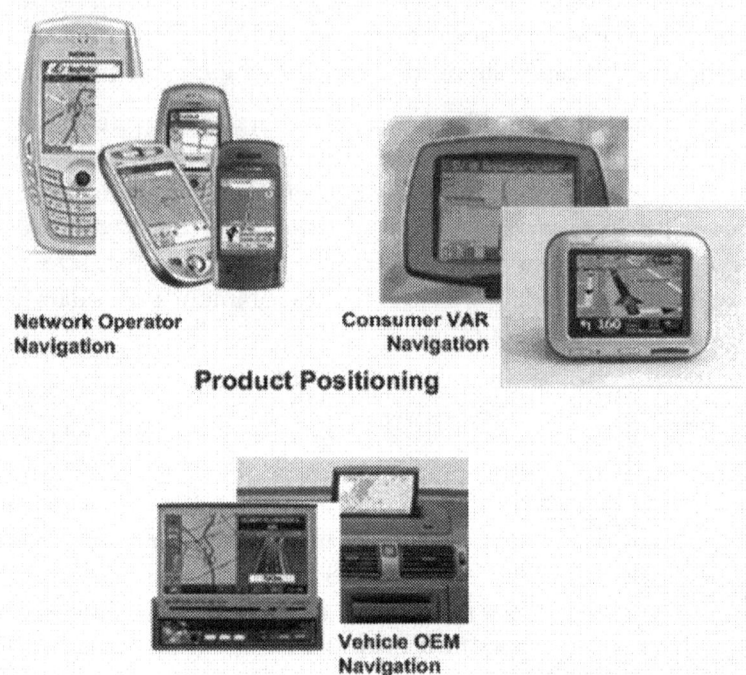

Photos taken by the author or permission to use photos granted by the system developers.

The bottom line is that there are navigation systems with traffic information to fit every budget. You can find your way to wherever you want to go, and you can be alerted about congestion and traffic problems along the way.

Figure 29: Portable navigation system statistics

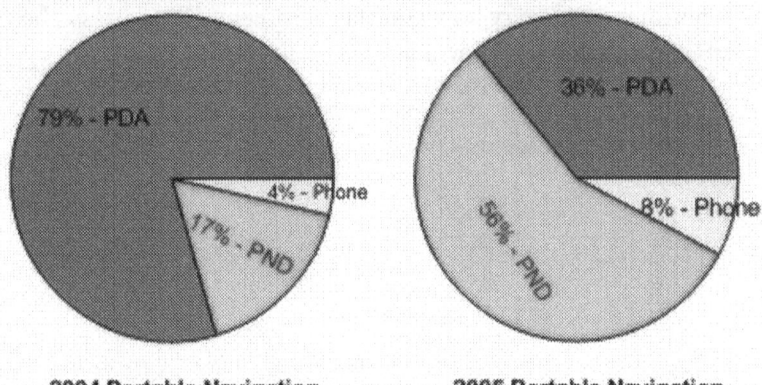

Portable Navigation Market - Device Market Share

2004 Portable Navigation — 79% - PDA, 4% - Phone, 17% - PND

2005 Portable Navigation — 36% - PDA, 8% - Phone, 56% - PND

Portable Navigation Market

Countries where sold		Outlets		Vendor Market Share	
Germany	28%	Consumer Elect	60%	TomTom	34%
UK	18	Car Accessories	5	Navigon	17
France	13	On-line Retail	8	Destinator	15
Benelux	13	Mobile Ph Retail	5	Navman	9
Italy	8	Mass Merchant	20	Garmin	6
Spain	7			ViaMoto	4
Other	13			Other	15

Source: Navteq Directions 2006

In 2007 there is almost no reason for anyone living in Europe or Japan, or in one of the fifty largest metropolitan areas in the U.S., to travel without the aid of a navigation system that provides real time traffic information. The cost of a well-functioning personal navigation device that delivers good directions and information on traffic problems had dropped to a price that made them affordable to most car owners, less than the price of a Big Mac for lunch every work day. There is therefore no reason to haplessly get stuck in traffic or to stay stuck in traffic if you are living in these areas.

You have a choice: You can load up your car with food and drink, personal hygiene and make-up articles, carry in your favorite music and audio books, and install a television for those times when you are stopped dead, all in order to make the painful journey in stop-and-go traffic bearable; or, you can buy a vehicle with a built-in navigation and traffic information system, or purchase an aftermarket system that you can move from car to car—make sure it includes the dynamic traffic information function—and make a positive plan to avoid wading out into the river of traffic when and where it is likely to be at its heaviest flow. Let the kids walk to school; separate transportation from recreation; shop locally; and, give commuting a rest every now and then. It seems like an easy choice to make, unless you truly do enjoy being stuck in traffic.

Acknowledgements

I am indebted to all who have helped me formulate my views that I have presented in this book, especially my wife, Britt Marie. We discussed all of the book's ideas, and she always encouraged me to frame my arguments in practical, rather than technical terms. My Mother, Mary Sena, was the first person to open my eyes to the Wal-Marting phenomenon when she pointed out that eighteen wheelers, rather than small vans, were delivering to the corner grocery store. Mike Dobson reviewed an early draft of the first few chapters and suggested that I change both the focus and the writing style. Approach it rationally, not emotionally, he said. I took his advice and started over. His periodic reviews and suggestions during the past two years were invaluable. Tom Crosby and Meng Lu also deserve thanks for their timely editorial comments. Tom's sharp journalistic eye spotted some weak points, and I took his suggestions for eliminating them. Meng had a look at the manuscript from a transportation engineering perspective and offered some very useful advice.

It took a few years to complete this book once I began writing it. I gathered research material and worked on the idea for a couple of years before putting the first words down on paper. From the time the idea for the book first entered my mind until the time I sent the manuscript to the publisher, I have seen my position change on some issues as new facts were presented and old information became discredited. For example, I went into this book as a firm believer in educational vouchers, and came out of it as a total supporter of small, local schools; I was indifferent to small, independently owned shops, but now I feel that they are one of the few remaining foundation stones on which we can re-build our communities; I was oblivious to the changes that had occurred in transportation and logistics and how they were affecting traffic flows, but now I see that we cannot resolve many of our traffic congestion issues in isolation of the economic realities of international trade and the movement of goods.

Writing a book, I have learned, is a journey. We decide to make the trip to a particular destination, and we start off with enough information to get us there. Along the way we meet new people and face new challenges. We gain experience from addressing the challenges, and we gain knowledge from all whom we meet. The journey is enriching in itself, and often we do not want it to end. Although the printing of these words on paper may seem to serve as a sign that the journey is over, for me it is the signal for the start of a new journey. I will now test the reactions received to this book and learn from them. What I learn I hope to share with you one day in the not too distant future.

Michael L. Sena
Åsa, Sweden

Beating Traffic: Time to Get Unstuck explores why, when and how congestion occurs, the part that you play in it, and what you and your family can do to reduce the negative effects of traffic congestion on your lives. Traffic congestion is not a pre-ordained state of affairs, and it is not you and I and the other car drivers who should have to live with it or bear the sole responsibility for fixing it. Decades of often well-meaning but horribly bad planning, and the conscious exploitation of real estate, have lead to most things being in the wrong place, too far from where everybody is and where everybody wants to go, so that the only way to get anywhere is to climb in the car and join the endless queues along the highways and local streets.

There are no quick fixes to the problem. Congestion charging is being promoted in many of the world's major cities as the ultimate solution. But turning the right of driving a car enjoyed by anyone into a privilege enjoyed only by those who can afford to pay the tolls charged by governments will only serve to increase the already wide chasm between the haves and the have-nots in the world. There are fairer and more equitable means to ration road space, and these are explored in this book.

While we are actively promoting real, long-term solutions to traffic congestion by supporting politicians who are serious about reconciling mobility and environmental sustainability, we can work on removing the inconvenience and irritation that it causes us on a daily basis. We can start by getting ourselves and our children out of our cars more often and onto the sidewalks, the bicycle paths and public transportation. We need to do our part to reduce harmful emissions by buying and driving responsibly, and we need to help stop the thousands of deaths and millions of injuries caused each year on our streets and highways by respecting the rules of the road and never driving under the influence of drugs or alcohol.

There are positive and concrete actions we can take to beat traffic, and you will read about them in *Beating Traffic*.

Michael Lawrence Sena is an internationally recognized expert in telematics, digital map databases, location-based services and navigation. He holds degrees in architecture and urban planning from Princeton University, is a registered architect, and practiced both in London and Boston before focusing on computer-aided mapping and geographic information. He has owned and run his consulting practice since 1983 with clients in Europe, North America and Asia in the mapping, automotive, software, system development, telematics service and database industries. During a four-year period, from 1993 through 1996, Michael worked for AB Volvo in Gothenburg, Sweden with responsibility for navigation, traffic information and fleet management data activities. He has served as an expert delegate to both the European CEN and international ISO standards committees on Digital Map Databases for intelligent transport systems. Michael lives and works on the west coast of Sweden with his wife, Britt Marie, and their cat, Iris. He has made *beating traffic* one of his major avocations.

Footnotes

[1] The *Economist* (December 17ᵗʰ 2005) reported that the American Institute of Medicine (IOM) has estimated around 16% of children aged 6-19 in the U.S. are now obese. That is three times the level in the 1960s. In both Europe and North America, public health agencies claim that it is the excess of high calorie and low nutrient foods marketed to children by the fast food chains that are to blame for putting children's long-term health at risk.

[2] *The Economist*, A Survey of Climate Change, Special Report, September 9ᵗʰ, 2006.

[3] Campaigns Team, London Chamber of Commerce and Industry. *The Third Retail Survey: The Impact of Congestion Charging on the Central London Retail Sector – Eighteen Months On.* January 2005

[4] New Statesman. *Give motorists a better deal – road tolls*, March 4, 2002.

[5] Shakespeare, William, *Julius Caesar*, Act III, Scene II.

[6] There are different names for fixed rail transit in different parts of the world. Trolleys are also called trams. The London Underground system is called 'The Underground', while in the U.S., such systems are often referred to as 'subways', or 'Ls' for elevated lines, or by their official names, like the 'BART' in San Francisco, the 'T' in Boston, or the 'Metro' in Washington, DC.

[7] Anthony Downs, "Urban Problems and Prospects" *Chapter 7, The Law of Peak-Hour Expressway Congestion* (Markham Publishing Company, 1970). Reprinted from Traffic Quarterly, Vol. 16, No. 3 (July, 1962), pp.393-409, by permission of the Eno Foundation for Transportation, Inc.

[8] By the terms of the Rail Passenger Service Act (1970), a National Railroad Passenger Corporation was created to operate virtually every intercity passenger rail line in the United States. Known as Amtrak, the quasi-public agency reduced the number of intercity passenger trains by one half in its first year of operation, retaining service only in areas of high-density travel.

[9] On June 29, 1956, President Eisenhower signed the Federal Aid-Highway Act of 1956, which authorized the interstate highway system (later formally named the Dwight D. Eisenhower System of Interstate and Defense Highways). The Act authorized 41,000 miles of high quality highways that were to tie the nation together. Later, congressional action increased the length to 42,500 miles and required super-highway standards for all interstate highways.

[10] Downs, op. cit.

[11] Downs, "Stuck in Traffic: Coping with Peak-Hour Traffic Congestion"(Brookings Institution Press Lincoln Institute of Land Policy, 1992).

[12] Anthony Downs: Testimony before the Committee on the Environment and Public Works, U.S. Senate (March 19, 2002).

[13] Traffic: America's great headache. *The Economist*, June 4ᵗʰ 2005. pp 44-45.

[14] Ibid., Two-thirds of Los Angeles residents surveyed consistently state that

traffic congestion—defined as freeway traffic moving at less than 35 miles per hour at peak commuting time—is "a big problem".

[15] Ibid.

[16] U.S. Department of Transportation, Federal Highway Administration, *National Household Travel Survey 2001*.

[17] The United Electrical, Radio and Machine Workers of America, *Fatigue and Shiftwork* (2005).

[18] UK Census Deprivation Study.

[19] Federal Reserve Bank of San Francisco. Community Investment Online, *Working Wheels*, 2005.

[20] The UK Commission for Integrated Transport, 2001.

[21] These are the vehicles that people drive for private use, and they include the full range of cars, sport utility vehicles, mini vans, and pick-up trucks sold in dealer showrooms around the world.

[22] The observation made in 1965 by Gordon Moore, co-founder of Intel, that the number of transistors per square inch on integrated circuits had doubled every year since the integrated circuit was invented. Moore predicted that this trend would continue for the foreseeable future. In subsequent years, the pace slowed down a bit, but data density has doubled approximately every 18 months, and this is the current definition of Moore's Law, which Moore himself has blessed. Most experts, including Moore himself, expect Moore's Law to hold for at least another two decades. (From *webopedia.com, the on-line encyclopedia dedicated to computer technology*, 2005).

[23] *The American College Dictionary*; Random House, New York, 1964.

[24] "What Englishman will give his mind to politics as long as he can afford to keep his motor car." **George Bernard Shaw**.

[25] Sweden is home to Saab Automobile and Volvo Cars, as well as to truck and bus manufacturers Scania and Volvo.

[26] Germany is home to Audi, BMW, DaimlerChrysler, Porsche, Volkswagen and GM Opel automobile manufacturers, and MAN and Daimler-Benz bus and truck manufacturers.

[27] Compare *automobile* to another word of Latin origin: *omnibus*, meaning "for all". From *omnibus* we get *bus*, the word for vehicles carrying many passengers.

[28] The American College Dictionary, op. cit.

[29] Elizabeth Moule and Stefanos Polyzoides, Los Angeles: Building the Polycentric Region; Congress for the New Urbanism (2005).

[30] Jane Jacobs. The Death and Life of Great American Cities. New York; Random House (1961)

[31] A French term for a dead end street which widens sufficiently at the end to permit an automobile to make a U-turn.

[32] U.S. Patent 1242872 - Self-serving store: Clarence Saunders, 1917.

[33] It was August 19, 1950 when the American Broadcasting Company (ABC) aired the first children's show. Favorites at that time were *Kukla, Fran and Ollie* and Buffalo Bob Smith's *Howdy Doody Time*.

[34] Jean Gottmann. Megalopolis: The Urbanized Northeastern Seaboard of the United States. New York; The Twentieth Century Fund (1961)

[35] It struck me as I wrote this list that all of these companies have disappeared. Boston is still there, and so is the traffic.

[36] Time Magazine. *Why Companies are Fleeing the Cities.* (April 26, 1971).

[37] Atkinson, Robert D. and Gottlieb, Paul D. The Metropolitan New Economy Index, PPI Policy Report, April 19, 2001.

[38] Moule and Polyzoides, op. cit.

[39] These words are copyright Jacqueline Steiner, and Bess Lomax-Hawes. The Kingston Trio version is copyright Capitol Records.

[40] **Con·ges·tion**: a state of excessive accumulation or overfilling or overcrowding. This is a general definition of a word that is used in many contexts. For example, a telecommunications network can be congested when there is not enough bandwidth to support the current traffic load. In medical terms, congestion refers to the presence of an abnormal amount of blood or other fluid in a vessel or organ. The origin of the word congest is from the Latin *congerere*, which means "heap up". In common speak, traffic congestion is referred to as a jam, a queue, a tie-up, a tailback.

[41] Road Information Program analysis of December 1998 report by the Federal Highway Administration.

[42] U.S. Federal Highway Administration, Office of Operations, 2000.

[43] U.S. Department of Transportation, Federal Highway Administration, Pulication No. FHWA-RD-00-067, pp. 8-14.

[44] Victoria Transport Policy Institute, Congestion Reduction Strategies: Identifying and Evaluating Strategies To Reduce Traffic Congestion; (June, 2004).

[45] Automatic Cruise Control is on the way.

[46] Document on Internet http://www.dft.gov.uk/stellent/groups/dft_transstats/documents/page/dft_transstats_032078.pdf

[47] Organisation for Economic Co-operation and Development - Brings together countries sharing the principles of the market economy, pluralist democracy and respect for human rights.

[48] World Health Organization, Deaths from Motor Vehicle Accidents in Selected Countries of the Americas, 1985-2001

[49] Compiled from various public sources.

[50] U.S. National Highway Traffic Safety Administration.

[51] 'Web site, inventors.about.com/library/inventors/blcar.htm, About, Inc., A part of the New York Times Company.

[52] GPS – Global Positioning System. See the last chapter for a full explanation of GPS technology.

[53] Heavy trucks are defined in the U.S. as vehicles that weigh over 14,000 pounds (6,363 kilograms), and are larger than seven feet (2.13 meters) tall, eight feet (2.44 meters) wide, and twenty-one feet (6.4 meters) long.

[54] In Europe, Heavy Commercial Vehicles are classified as those over 16 tons (35,200 pounds).

[55] Larger and heavier vehicles tend to require more road space and are slower to accelerate, and so cause more traffic congestion than smaller, lighter vehicles. The relative congestion impact of different vehicles is measured in terms of "Passenger Car Equivalents" or PCEs. Large trucks and buses tend to have 1.5-4 PCEs, depending on roadway conditions, as shown in Table 3, and even more through intersections or under stop-and-go driving conditions. A large SUV imposes 1.4 PCEs and a van 1.3 PCEs when traveling through an intersection. (Raheel Shabih and Kara M. Kockelman, *Effect of Vehicle Type on the Capacity of Signalized Intersections: The Case of Light-Duty Trucks*, University of Texas at Austin, 1999).

[56] Many companies have tried to copy Toyota's famous production system--but without success. Part of the reason why, says the author, is that imitators fail to recognize the underlying principles of the Toyota Production System (TPS), focusing instead on specific tools and practices. Steven J. Spear, *Learning to Lead at Toyota* (Harvard Business Review, May 1, 2004).

[57] In the UK, this type of vehicle is called an "estate car".

[58] In the summer of 2006, Wal-Mart sold its eighty-five stores in Germany and left the market, taking a one billion charge against its earnings. The reason: its local competitors, especially Metro, were better at the Wal-Marting game than Wal-Mart.

[59] Wal-Mart had been 10% larger than its closest competitors, BP, until the price of oil skyrocketed in 2005 to over $50 per barrel of crude. This windfall pushed the sales of oil giants BP and Royal Dutch Shell Group to levels close to those of Wal-Mart, and Exxon Mobil past it. In one year, Royal Dutch/Shell Group sales doubled, from $133 billion to $269 billion.

[60] In 2004, Swift had 18,500 trucks; Schneider had 15,000 trucks; and J.B. Hunt had 6,000 trucks. Wal-Mart was Swift's largest customer, accounting for 15% of total sales.

[61] Some stores have evolved into warehouse shopping facilities with a minimum of frills and no service, and with goods piled up on pallets for picking by the customers. High shelves in the same space where he customer is shopping are packed with goods that would normally be in a storage facility out of the customers sight.

[62] Arkansas ranks 49th out of 50 in Median Household Income: 1999-2001; 1st in Persons Below Poverty Line: 2000-2002; and 46th in Average Annual Pay: 2002. *U.S. Census Bureau, Statistical Abstract of the United States* (2004).

[63] This area of the U.S. has another advantage. It has a large supply of abandoned coal mines that can be filled with the earth that is excavated to form the pits for burying the waste.

[64] Carbon monoxide emissions for trucks (105 grams per ton of transport per kilometer) are three times greater than for rail (35 g/tkm), and over five times greater than for shipping (18 g/tkm).

[65] Walberg, H. *On Local Control: Is Bigger Better? Source Book on School and District Size, Cost and Quality.* Hubert H. Humphrey Institute of Public Affairs, University of Minnesota, Minneapolis, 1992, pp. 118-134.

[66] Nationwide Personal Transportation Study. *Transportation Characteristics of School Children, Report No. 4*. Federal Highway Administration, Washington, D.C., 1972.

[67] Dellinger, A., and Staunton, C. Barriers to Children Walking and Bicycling to School: United States, 1999, *Morbidity and Mortality Weekly Report*, Vol. 51, No.32, 2002.

[68] Op. Cit., Nationwide Personal Transportation Study.

[69] Public schools in the United States could be racially segregated until 1954, meaning that Black and White children could be forced to attend different schools. A Supreme Court ruling from 1892, Plessy v. Ferguson, legitimized these children's "separate, but equal" educations.

[70] House of Commons Transport Select Committee. *Inquiry into Home to School Transport*, A memorandum from the Sutton Trust, February 2004.

[71] Macket (2001). U.K. Department for Transport study.

[72] Sena, Michael L., "Simulation of Alternative Educational Strategies", *Town Planning Review*, Vol. 47, No. 2, April 1976.

[73] In 1980, 6% of American children were overweight. In 2004, 16% were considered obese.

[74] The first child car seats were invented in 1921, following the introduction of Henry Ford's Model T, however, they were very different from today's car seat. The earliest versions were essentially sacks with drawstring attached to the back seat. In 1978, Tennessee became the first American State to require child safety seat use.

[75] Definition provided by WordReference.com, 2005.

[76] Auto and Truck Seasonal Adjustment, United States Department of Commerce, Bureau of Economic Analysis (BEA), Washington, D.C., August 3, 2005.

[77] Fine, Judylaine, *Conquering Back Pain*. New York: Prentice Hall Press, 1987.

[78] Hawkins, Steven, A Brief History of Time. London: Bantam Press, 1998. He has proof that the chicken had to come first.

[79] A *medina* is the old section of an Arab city in North Africa. When capitalized, *Medina* is originally the name of a city in western Saudi Arabia.

[80] Shopping is the act of exchanging, or considering the exchange, of something that you have of value, like money, for something of value that someone else has, like a product or service. A **shop** is where the transactions occur. There are barber shops, butcher shops, candy shops, grocery shops and many other kinds of shops. Another word for a *shop* that is mostly used in the U.S. is **store**, which comes from the Middle English word *stor*, to "supply". The word *shop* is a shortening of the Old French word *eschoppe*, which means "lean-to booth". A lean-to is a sort of tent open on one side. Lean-to shops have been replaced by permanent structures in most of the industrialized western countries, although there are still market days in many European cities with temporary stalls reminiscent of the earliest shops.

[81] Mitchell, S. and Milchen, J., *Littering the West with Dead Malls and Vacant*

Superstores. Amiba Articles, 2001.

[82] Ibid. Chain stores are multiplying at a staggering pace and have created a glut of retail space. Wal-Mart is one of the worst offenders with nearly 400 of its stores—many less than a decade old—sitting empty. It abandons the old stores to build new and larger facilities in the same vicinity.

[83] In January 2006, China reported that its overall trade surplus stood at $102 billion in 2005, up from $32 billion in 2004. China's surplus with the United States was $114.7 billion. A trade surplus is the difference between what a country sells to other countries and what it purchases from other countries.

[84] The European Union had fifteen members between 1995 and 2004. The fifteen were: Germany, France, United Kingdom, Italy, The Netherlands, Belgium, Luxembourg, Spain, Portugal, Sweden, Finland, Austria, Republic of Ireland, Denmark, Greece. The ten new states added in 2004 were: Cyprus, Czech Republic, Estonia, Hungary, Latvia, Lithuania, Malta, Poland, Slovakia, Slovenia.

[85] The Economist, Material Fitness (pp. 65-66), February 25th 2006.

[86] The Economist, A Survey of the World Economy, Special Report, September 16th, 2006.

[87] *The Wall Street Journal*, "For more Americans life begins at 5 a.m.". (Monday, March 27, 2006).

[88] Breitenberger, Susanne, *Necessary Penetration Rates of Probe Vehicles*, Paper presented at ITS World Congress, San Francisco (2005). Type A roads are the highest category of road, and Type F the lowest.

[89] *(For* Greyhound Bus, *Grey Advertising, 1957).*

[90] The American English Dictionary, Random House, Inc. 1964.

[91] Lanciani, Rodolfo, Ancient Rome in the Light of Recent Discoveries. Boston and New York: Houghto, Mifflin and Company, 1898. Reproduced on the World Wide Web by William P. Thayer.

[92] *Octroi* is a local tax levied on certain articles, such as foodstuffs, on their admission into a city. Also, the place at which the tax is collected or the person collecting it. Derived from the French, *octroyer*, "to grant". (The American College Dictionary, op. cit.)

[93] In 2005, the U.S. budget deficit was $318 billion. In 2001, when George W. Bush took over as President, he inherited a $526 billion surplus by the previous President, Bill Clinton.

[94] From Florilegium Urbanum, Copyright Stephen Alsford (2001-2006); www.trytel.com. *Cordwain* was a soft, fine-grained leather originally produced in Cordoba, originally using goat-skin, and tawed with alum; it was particularly used in shoemaking. *Godelmynges* appear to have been the same, except using the skins of young animals; It is suggested this term derived from Godalming, Surrey.

[95] Extracts reproduced here from the Parliamentary Committee Report, *Statement of the Toll Reform Committee*, were obtained from a web site that reproduced text from The Illustrated London News, June 6, 1857.

[96] www.stockholmsforsoket.se – The Stockholm Trials cost is 1.8 billion

Swedish kronor.

[97] *The Intelligent Highway*, "Stockholm Congestion Charging Produces 25% Drop in Traffic", pp. 4-5, April 1, 2006.

[98] Such exceptions were under consideration by the London Mayor's office at the time of this writing.

[99] The report was produced by consulting company Transek.

[100] Göteborgs Posten, August 17, 2006.

[101] The Intelligent Highway, Highway Agency Sees Future Of In-Car Information Services, May 1, 2006, (pp. 8-9).

[102] Morse Code is a system of dots, dashes and spaces or corresponding sounds, used in telegraphy and signaling to represent the letters of the alphabet, numerals, etc. It is named after its inventor, Samuel Finley Breese Morse (1791-1872).

[103] Definitions for probe and float are from The World Book Encyclopedia Dictionary, Doubleday & Company, Inc., 1963.

[104] A Brief History of the Numbering System of UK Roads; James Bufford: A 1919 Act of Parliament provided funding for roads to the Ministry of Transport. By 1921, the MoT had identified the system for England and classified 97 main A roads with one and two digit numbers. The formal classification was published in 1923 by HMSO in a booklet, and the Ordnance Survey of Great Britain published a series of maps for MoT which included the road numbers. The numbers were placed on road signs at the same time.

[105] From Wikipedia, the free encyclopedia. from "http://en.wikipedia.org/wiki/CD-ROM"

[106] GNSS is a generic term to denote all types of satellite positioning systems.

[107] The **speed of light** in a vacuum is defined to be exactly 299,792,458 meters per second (or 1,079,252,848.8 km/h, which is approximately 186,282.397 miles per second, or 670,616,629.4 miles per hour). This value is denoted by the letter c, reputedly from the Latin celeritas, "speed", and also known as **Einstein's constant**. (Wikipedia, The Free Encyclopedia)

[108] One of the main reasons GPS started out as only a minor part of the positioning process was that the positional accuracy of the signal was around +/-100 meters. This was not sufficient to allow a moving vehicle to be reliably placed on a road, especially in tightly packed urban areas. In May, 2005, the U.S. removed Selective Availability, and the positional accuracy immediately increased to less than +/-10 meters.

[109] Sena, Michael L., "Computer Mapping for Publication", *Computer Graphics World*, Vol. 6, Number 7, (July 1983), pp.68-76.

www.ingramcontent.com/pod-product-compliance
Lightning Source LLC
Chambersburg PA
CBHW061314280526
45784CB00002B/984